Springer Undergraduate Mathematics Series

Advisory Board

Other books in this series

Karin Erdmann and Mark J. Wildon

Introduction to Lie Algebras

 Springer

Karin Erdmann
Mathematical Institute
24–29 St Giles'
Oxford OX1 3LB
UK
Erdmann@maths.ox.ac.uk

Mark J. Wildon
Mathematical Institute
24–29 St Giles'
Oxford OX1 3LB
UK
wildon@maths.ox.ac.uk

British Library Cataloguing in Publication Data
A catalogue record for this book is available from the British Library

Library of Congress Control Number: 2005937687

Springer Undergraduate Mathematics Series ISSN 1615-2085

ISBN 978-1-84628-040-5

Printed on acid-free paper

9 8 7 6 5 4 3 2   (corrected as of 2nd printing, 2007)

Springer Science+Business Media
springer.com

Preface

Lie theory has its roots in the work of Sophus Lie, who studied certain transformation groups that are now called Lie groups. His work led to the discovery of Lie algebras. By now, both Lie groups and Lie algebras have become essential to many parts of mathematics and theoretical physics. In the meantime, Lie algebras have become a central object of interest in their own right, not least because of their description by the Serre relations, whose generalisations have been very important.

This text aims to give a very basic algebraic introduction to Lie algebras. We begin here by mentioning that "Lie" should be pronounced "lee". The only prerequisite is some linear algebra; we try throughout to be as simple as possible, and make no attempt at full generality. We start with fundamental concepts, including ideals and homomorphisms. A section on Lie algebras of small dimension provides a useful source of examples. We then define solvable and simple Lie algebras and give a rough strategy towards the classification of the finite-dimensional complex Lie algebras. The next chapters discuss Engel's Theorem, Lie's Theorem, and Cartan's Criteria and introduce some representation theory.

We then describe the root space decomposition of a semisimple Lie algebra and introduce Dynkin diagrams to classify the possible root systems. To practice these ideas, we find the root space decompositions of the classical Lie algebras. We then outline the remarkable classification of the finite-dimensional simple Lie algebras over the complex numbers.

The final chapter is a survey on further directions. In the first part, we introduce the universal enveloping algebra of a Lie algebra and look in more detail at representations of Lie algebras. We then look at the Serre relations and their generalisations to Kac–Moody Lie algebras and quantum groups and describe the Lie ring associated to a group. In fact, Dynkin diagrams and the classification of the finite-dimensional complex semisimple Lie algebras have had a far-reaching influence on modern mathematics; we end by giving an illustration of this.

In Appendix A, we give a summary of the basic linear and bilinear algebra we need. Some technical proofs are deferred to Appendices B, C, and D. In Appendix E, we give answers to some selected exercises. We do, however, encourage the reader to make a thorough unaided attempt at these exercises: it is only when treated in this way that they will be of any benefit. Exercises are marked † if an answer may be found in Appendix E and ⋆ if they are either somewhat harder than average or go beyond the usual scope of the text.

University of Oxford   Karin Erdmann
January 2006   Mark Wildon

Contents

1
Introduction

We begin by defining Lie algebras and giving a collection of typical examples to which we shall refer throughout this book. The remaining sections in this chapter introduce the basic vocabulary of Lie algebras. The reader is reminded that the prerequisite linear and bilinear algebra is summarised in Appendix A.

1.1 Definition of Lie Algebras

Let F be a field. A *Lie algebra* over F is an F-vector space L, together with a bilinear map, the *Lie bracket*

$$L \times L \to L, \quad (x, y) \mapsto [x, y],$$

satisfying the following properties:

$$[x, x] = 0 \quad \text{for all } x \in L, \tag{L1}$$

$$[x, [y, z]] + [y, [z, x]] + [z, [x, y]] = 0 \quad \text{for all } x, y, z \in L. \tag{L2}$$

The Lie bracket $[x, y]$ is often referred to as the *commutator* of x and y. Condition (L2) is known as the *Jacobi identity*. As the Lie bracket $[-, -]$ is bilinear, we have

$$0 = [x + y, x + y] = [x, x] + [x, y] + [y, x] + [y, y] = [x, y] + [y, x].$$

Hence condition (L1) implies

$$[x, y] = -[y, x] \quad \text{for all } x, y \in L. \tag{L1$'$}$$

If the field F does not have characteristic 2, then putting $x = y$ in (L1′) shows that (L1′) implies (L1).

Unless specifically stated otherwise, all Lie algebras in this book should be taken to be finite-dimensional. (In Chapter 15, we give a brief introduction to the more subtle theory of infinite-dimensional Lie algebras.)

Exercise 1.1

(i) Show that $[v, 0] = 0 = [0, v]$ for all $v \in L$.

(ii) Suppose that $x, y \in L$ satisfy $[x, y] \neq 0$. Show that x and y are linearly independent over F.

1.2 Some Examples

(1) Let $F = \mathbf{R}$. The vector product $(x, y) \mapsto x \wedge y$ defines the structure of a Lie algebra on $\mathbf{R}^3$. We denote this Lie algebra by $\mathbf{R}^3_\wedge$. Explicitly, if $x = (x_1, x_2, x_3)$ and $y = (y_1, y_2, y_3)$, then

$$x \wedge y = (x_2 y_3 - x_3 y_2, x_3 y_1 - x_1 y_3, x_1 y_2 - x_2 y_1).$$

Exercise 1.2

Convince yourself that $\wedge$ is bilinear. Then check that the Jacobi identity holds. *Hint*: If $x \cdot y$ denotes the dot product of the vectors $x, y \in \mathbf{R}^3$, then

$$x \wedge (y \wedge z) = (x \cdot z)y - (x \cdot y)z \quad \text{for all } x, y, z \in \mathbf{R}^3.$$

(2) Any vector space V has a Lie bracket defined by $[x, y] = 0$ for all $x, y \in V$. This is the *abelian* Lie algebra structure on V. In particular, the field F may be regarded as a 1-dimensional abelian Lie algebra.

(3) Suppose that V is a finite-dimensional vector space over F. Write $\mathsf{gl}(V)$ for the set of all linear maps from V to V. This is again a vector space over F, and it becomes a Lie algebra, known as the *general linear algebra*, if we define the Lie bracket $[-, -]$ by

$$[x, y] := x \circ y - y \circ x \quad \text{for } x, y \in \mathsf{gl}(V),$$

where $\circ$ denotes the composition of maps.

Exercise 1.3

Check that the Jacobi identity holds. (This exercise is famous as one that every mathematician should do at least once in her life.)

(3′) Here is a matrix version. Write $\mathsf{gl}(n, F)$ for the vector space of all $n \times n$ matrices over F with the Lie bracket defined by

$$[x, y] := xy - yx,$$

where xy is the usual product of the matrices x and y.

As a vector space, $\mathsf{gl}(n, F)$ has a basis consisting of the *matrix units* e_{ij} for $1 \leq i, j \leq n$. Here e_{ij} is the $n \times n$ matrix which has a 1 in the ij-th position and all other entries are 0. We leave it as an exercise to check that

$$[e_{ij}, e_{kl}] = \delta_{jk} e_{il} - \delta_{il} e_{kj},$$

where δ is the Kronecker delta, defined by $\delta_{ij} = 1$ if $i = j$ and $\delta_{ij} = 0$ otherwise. This formula can often be useful when calculating in $\mathsf{gl}(n, F)$.

(4) Recall that the trace of a square matrix is the sum of its diagonal entries. Let $\mathsf{sl}(n, F)$ be the subspace of $\mathsf{gl}(n, F)$ consisting of all matrices of trace 0. For arbitrary square matrices x and y, the matrix $xy - yx$ has trace 0, so $[x, y] = xy - yx$ defines a Lie algebra structure on $\mathsf{sl}(n, F)$: properties (L1) and (L2) are inherited from $\mathsf{gl}(n, F)$. This Lie algebra is known as the *special linear algebra*. As a vector space, $\mathsf{sl}(n, F)$ has a basis consisting of the e_{ij} for $i \neq j$ together with $e_{ii} - e_{i+1,i+1}$ for $1 \leq i < n$.

(5) Let $\mathsf{b}(n, F)$ be the upper triangular matrices in $\mathsf{gl}(n, F)$. (A matrix x is said to be upper triangular if $x_{ij} = 0$ whenever $i > j$.) This is a Lie algebra with the same Lie bracket as $\mathsf{gl}(n, F)$.

Similarly, let $\mathsf{n}(n, F)$ be the strictly upper triangular matrices in $\mathsf{gl}(n, F)$. (A matrix x is said to be strictly upper triangular if $x_{ij} = 0$ whenever $i \geq j$.) Again this is a Lie algebra with the same Lie bracket as $\mathsf{gl}(n, F)$.

Exercise 1.4

Check the assertions in (5).

1.3 Subalgebras and Ideals

The last two examples suggest that, given a Lie algebra L, we might define a *Lie subalgebra* of L to be a vector subspace $K \subseteq L$ such that

$$[x, y] \in K \quad \text{for all } x, y \in K.$$

Lie subalgebras are easily seen to be Lie algebras in their own right. In Examples (4) and (5) above we saw three Lie subalgebras of $\mathsf{gl}(n, F)$.

We also define an *ideal* of a Lie algebra L to be a subspace I of L such that

$$[x,y] \in I \quad \text{for all } x \in L, \ y \in I.$$

By (L1'), $[x,y] = -[y,x]$, so we do not need to distinguish between left and right ideals. For example, $\mathsf{sl}(n,F)$ is an ideal of $\mathsf{gl}(n,F)$, and $\mathsf{n}(n,F)$ is an ideal of $\mathsf{b}(n,F)$.

An ideal is always a subalgebra. On the other hand, a subalgebra need not be an ideal. For example, $\mathsf{b}(n,F)$ is a subalgebra of $\mathsf{gl}(n,F)$, but provided $n \geq 2$, it is not an ideal. To see this, note that $e_{11} \in \mathsf{b}(n,F)$ and $e_{21} \in \mathsf{gl}(n,F)$. However, $[e_{21}, e_{11}] = e_{21} \notin \mathsf{b}(n,F)$.

The Lie algebra L is itself an ideal of L. At the other extreme, $\{0\}$ is an ideal of L. We call these the *trivial ideals* of L. An important example of an ideal which frequently is non-trivial is the *centre* of L, defined by

$$Z(L) := \{x \in L : [x,y] = 0 \text{ for all } y \in L\}.$$

We know precisely when $L = Z(L)$ as this is the case if and only if L is abelian. On the other hand, it might take some work to decide whether or not $Z(L) = \{0\}$.

Exercise 1.5

Find $Z(L)$ when $L = \mathsf{sl}(2,F)$. You should find that the answer depends on the characteristic of F.

1.4 Homomorphisms

If L_1 and L_2 are Lie algebras over a field F, then we say that a map $\varphi : L_1 \to L_2$ is a *homomorphism* if φ is a linear map and

$$\varphi([x,y]) = [\varphi(x), \varphi(y)] \quad \text{for all } x, y \in L_1.$$

Notice that in this equation the first Lie bracket is taken in L_1 and the second Lie bracket is taken in L_2. We say that φ is an *isomorphism* if φ is also bijective.

An extremely important homomorphism is the *adjoint homomorphism* . If L is a Lie algebra, we define

$$\mathrm{ad} : L \to \mathsf{gl}(L)$$

by $(\mathrm{ad}\,x)(y) := [x,y]$ for $x, y \in L$. It follows from the bilinearity of the Lie bracket that the map $\mathrm{ad}\,x$ is linear for each $x \in L$. For the same reason, the map $x \mapsto \mathrm{ad}\,x$ is itself linear. So to show that ad is a homomorphism, all we need to check is that

$$\mathrm{ad}([x,y]) = \mathrm{ad}\,x \circ \mathrm{ad}\,y - \mathrm{ad}\,y \circ \mathrm{ad}\,x \quad \text{for all } x, y \in L;$$

this turns out to be equivalent to the Jacobi identity. The kernel of ad is the centre of L.

Exercise 1.6

Show that if $\varphi : L_1 \to L_2$ is a homomorphism, then the kernel of φ, ker φ, is an ideal of L_1, and the image of φ, im φ, is a Lie subalgebra of L_2.

Remark 1.1

Whenever one has a mathematical object, such as a vector space, group, or Lie algebra, one has attendant homomorphisms. Such maps are of interest precisely because they are structure preserving — *homo*, same; *morphos*, shape. For example, working with vector spaces, if we add two vectors, and then apply a homomorphism of vector spaces (also known as a linear map), the result should be the same as if we had first applied the homomorphism, and then added the image vectors.

Given a class of mathematical objects one can (with some thought) work out what the relevant homomorphisms should be. Studying these homomorphisms gives one important information about the structures of the objects concerned. A common aim is to classify all objects of a given type; from this point of view, we regard isomorphic objects as essentially the same. For example, two vector spaces over the same field are isomorphic if and only if they have the same dimension.

1.5 Algebras

An *algebra* over a field F is a vector space A over F together with a bilinear map,

$$A \times A \to A, \quad (x, y) \mapsto xy.$$

We say that xy is the *product* of x and y. Usually one studies algebras where the product satisfies some further properties. In particular, Lie algebras are the algebras satisfying identities (L1) and (L2). (And in this case we write the product xy as $[x, y]$.)

The algebra A is said to be *associative* if

$$(xy)z = x(yz) \quad \text{for all } x, y, z \in A$$

and *unital* if there is an element 1_A in A such that $1_A x = x = x1_A$ for all non-zero elements of A.

For example, $\mathsf{gl}(V)$, the vector space of linear transformations of the vector space V, has the structure of a unital associative algebra where the product is given by the composition of maps. The identity transformation is the identity element in this algebra. Likewise $\mathsf{gl}(n, F)$, the set of $n \times n$ matrices over F, is a unital associative algebra with respect to matrix multiplication.

Apart from Lie algebras, most algebras one meets tend to be both associative and unital. It is important not to get confused between these two types of algebras. One way to emphasise the distinction, which we have adopted, is to always write the product in a Lie algebra with square brackets.

Exercise 1.7

Let L be a Lie algebra. Show that the Lie bracket is associative, that is, $[x, [y, z]] = [[x, y], z]$ for all $x, y, z \in L$, if and only if for all $a, b \in L$ the commutator $[a, b]$ lies in $Z(L)$.

If A is an associative algebra over F, then we define a new bilinear operation $[-, -]$ on A by

$$[a, b] := ab - ba \quad \text{for all } a, b \in A.$$

Then A together with $[-, -]$ is a Lie algebra; this is not hard to prove. The Lie algebras $\mathsf{gl}(V)$ and $\mathsf{gl}(n, F)$ are special cases of this construction. In fact, if you did Exercise 1.3, then you will already have proved that the product $[-, -]$ satisfies the Jacobi identity.

1.6 Derivations

Let A be an algebra over a field F. A *derivation* of A is an F-linear map $D : A \to A$ such that

$$D(ab) = aD(b) + D(a)b \quad \text{for all } a, b \in A.$$

Let $\text{Der}\,A$ be the set of derivations of A. This set is closed under addition and scalar multiplication and contains the zero map. Hence $\text{Der}\,A$ is a vector subspace of $\mathsf{gl}(A)$. Moreover, $\text{Der}\,A$ is a Lie subalgebra of $\mathsf{gl}(A)$, for by part (i) of the following exercise, if D and E are derivations then so is $[D, E]$.

Exercise 1.8

Let D and E be derivations of an algebra A.

(i) Show that $[D, E] = D \circ E - E \circ D$ is also a derivation.

(ii) Show that $D \circ E$ need not be a derivation. (The following example may be helpful.)

Example 1.2

(1) Let $A = C^\infty \mathbf{R}$ be the vector space of all infinitely differentiable functions $\mathbf{R} \to \mathbf{R}$. For $f, g \in A$, we define the product fg by pointwise multiplication: $(fg)(x) = f(x)g(x)$. With this definition, A is an associative algebra. The usual derivative, $Df = f'$, is a derivation of A since by the product rule

$$D(fg) = (fg)' = f'g + fg' = (Df)g + f(Dg).$$

(2) Let L be a Lie algebra and let $x \in L$. The map $\operatorname{ad} x : L \to L$ is a derivation of L since by the Jacobi identity

$$(\operatorname{ad} x)[y, z] = [x, [y, z]] = [[x, y], z] + [y, [x, z]] = [(\operatorname{ad} x)y, z] + [y, (\operatorname{ad} x)z]$$

for all $y, z \in L$.

1.7 Structure Constants

If L is a Lie algebra over a field F with basis $(x_1, \ldots, x_n)$, then $[-, -]$ is completely determined by the products $[x_i, x_j]$. We define scalars $a_{ij}^k \in F$ such that

$$[x_i, x_j] = \sum_{k=1}^n a_{ij}^k x_k.$$

The a_{ij}^k are the *structure constants* of L with respect to this basis. We emphasise that the a_{ij}^k depend on the choice of basis of L: Different bases will in general give different structure constants.

By (L1) and its corollary (L1'), $[x_i, x_i] = 0$ for all i and $[x_i, x_j] = -[x_j, x_i]$ for all i and j. So it is sufficient to know the structure constants a_{ij}^k for $1 \leq i < j \leq n$.

Exercise 1.9

Let L_1 and L_2 be Lie algebras. Show that L_1 is isomorphic to L_2 if and only if there is a basis B_1 of L_1 and a basis B_2 of L_2 such that the structure constants of L_1 with respect to B_1 are equal to the structure constants of L_2 with respect to B_2.

Exercise 1.10

Let L be a Lie algebra with basis $(x_1, \ldots, x_n)$. What condition does the Jacobi identity impose on the structure constants a_{ij}^k?

EXERCISES

1.11.† Let L_1 and L_2 be two abelian Lie algebras. Show that L_1 and L_2 are isomorphic if and only if they have the same dimension.

1.12.† Find the structure constants of $\mathsf{sl}(2, F)$ with respect to the basis given by the matrices

$$e = \begin{pmatrix} 0 & 1 \\ 0 & 0 \end{pmatrix}, \ f = \begin{pmatrix} 0 & 0 \\ 1 & 0 \end{pmatrix}, \ h = \begin{pmatrix} 1 & 0 \\ 0 & -1 \end{pmatrix}.$$

1.13. Prove that $\mathsf{sl}(2, \mathbf{C})$ has no non-trivial ideals.

1.14.† Let L be the 3-dimensional *complex* Lie algebra with basis (x, y, z) and Lie bracket defined by

$$[x, y] = z, \ [y, z] = x, \ [z, x] = y.$$

(Here L is the "complexification" of the 3-dimensional real Lie algebra $\mathbf{R}_\wedge^3$.)

(i) Show that L is isomorphic to the Lie subalgebra of $\mathsf{gl}(3, \mathbf{C})$ consisting of all 3×3 antisymmetric matrices with entries in $\mathbf{C}$.

(ii) Find an explicit isomorphism $\mathsf{sl}(2, \mathbf{C}) \cong L$.

1.15. Let S be an $n \times n$ matrix with entries in a field F. Define

$$\mathsf{gl}_S(n, F) = \{x \in \mathsf{gl}(n, F) : x^t S = -Sx\}.$$

(i) Show that $\mathsf{gl}_S(n, F)$ is a Lie subalgebra of $\mathsf{gl}(n, F)$.

(ii) Find $\mathsf{gl}_S(2, \mathbf{R})$ if $S = \begin{pmatrix} 0 & 1 \\ 0 & 0 \end{pmatrix}$.

(iii) Does there exist a matrix S such that $\mathsf{gl}_S(2, \mathbf{R})$ is equal to the set of all diagonal matrices in $\mathsf{gl}(2, \mathbf{R})$?

(iv) Find a matrix S such that $\mathsf{gl}_S(3, \mathbf{R})$ is isomorphic to the Lie algebra $\mathbf{R}_\wedge^3$ defined in §1.2, Example 1.

Hint: Part (i) of Exercise 1.14 is relevant.

1.16.† Show, by giving an example, that if F is a field of characteristic 2, there are algebras over F which satisfy (L1′) and (L2) but are not Lie algebras.

1.17. Let V be an n-dimensional complex vector space and let $L = \mathsf{gl}(V)$. Suppose that $x \in L$ is diagonalisable, with eigenvalues $\lambda_1, \ldots, \lambda_n$. Show that $\operatorname{ad} x \in \mathsf{gl}(L)$ is also diagonalisable and that its eigenvalues are $\lambda_i - \lambda_j$ for $1 \le i, j \le n$.

1.18.   Let L be a Lie algebra. We saw in §1.6, Example 1.2(2) that the maps
         $\operatorname{ad} x : L \to L$ for $x \in L$ are derivations of L; these are known as *inner
         derivations*. Show that if $\operatorname{IDer} L$ is the set of inner derivations of L,
         then $\operatorname{IDer} L$ is an ideal of $\operatorname{Der} L$.

1.19.   Let A be an algebra and let $\delta : A \to A$ be a derivation. Prove that δ
         satisfies the Leibniz rule

$$\delta^n(xy) = \sum_{r=0}^{n} \binom{n}{r} \delta^r(x) \delta^{n-r}(y) \quad \text{for all } x, y \in A.$$

2

Ideals and Homomorphisms

In this chapter we explore some of the constructions in which ideals are involved. We shall see that in the theory of Lie algebras ideals play a role similar to that played by normal subgroups in the theory of groups. For example, we saw in Exercise 1.6 that the kernel of a Lie algebra homomorphism is an ideal, just as the kernel of a group homomorphism is a normal subgroup.

2.1 Constructions with Ideals

Suppose that I and J are ideals of a Lie algebra L. There are several ways we can construct new ideals from I and J. First we shall show that $I \cap J$ is an ideal of L. We know that $I \cap J$ is a subspace of L, so all we need check is that if $x \in L$ and $y \in I \cap J$, then $[x, y] \in I \cap J$: This follows at once as I and J are ideals.

Exercise 2.1

Show that $I + J$ is an ideal of L where

$$I + J := \{x + y : x \in I, \ y \in J\}.$$

We can also define a product of ideals. Let

$$[I, J] := \mathrm{Span}\{[x, y] : x \in I, y \in J\}.$$

We claim that $[I, J]$ is an ideal of L. Firstly, it is by definition a subspace. Secondly, if $x \in I$, $y \in J$, and $u \in L$, then the Jacobi identity gives

$$[u, [x, y]] = [x, [u, y]] + [[u, x], y].$$

Here $[u, y] \in J$ as J is an ideal, so $[x, [u, y]] \in [I, J]$. Similarly, $[[u, x], y] \in [I, J]$. Therefore their sum belongs to $[I, J]$.

A general element t of $[I, J]$ is a linear combination of brackets $[x, y]$ with $x \in I$, $y \in J$, say $t = \sum c_i [x_i, y_i]$, where the c_i are scalars and $x_i \in I$ and $y_i \in J$. Then, for any $u \in L$, we have

$$[u, t] = \left[u, \sum c_i [x_i, y_i] \right] = \sum c_i [u, [x_i, y_i]],$$

where $[u, [x_i, y_i]] \in [I, J]$ as shown above. Hence $[u, t] \in [I, J]$ and so $[I, J]$ is an ideal of L.

Remark 2.1

It is necessary to define $[I, J]$ to be the *span* of the commutators of elements of I and J rather than just the *set* of such commutators. See Exercise 2.14 below for an example where the set of commutators is not itself an ideal.

An important example of this construction occurs when we take $I = J = L$. We write L' for $[L, L]$: Despite being an ideal of L, L' is usually known as the *derived algebra* of L'. The term *commutator algebra* is also sometimes used.

Exercise 2.2

Show that $\mathsf{sl}(2, \mathbf{C})' = \mathsf{sl}(2, \mathbf{C})$.

2.2 Quotient Algebras

If I is an ideal of the Lie algebra L, then I is in particular a subspace of L, and so we may consider the cosets $z + I = \{z + x : x \in I\}$ for $z \in L$ and the quotient vector space

$$L/I = \{z + I : z \in L\}.$$

We review the vector space structure of L/I in Appendix A. We claim that a Lie bracket on L/I may be defined by

$$[w + I, z + I] := [w, z] + I \quad \text{for } w, z \in L.$$

Here the bracket on the right-hand side is the Lie bracket in L. To be sure that the Lie bracket on L/I is well-defined, we must check that $[w, z] + I$

depends only on the cosets containing w and z and not on the particular coset representatives w and z. Suppose $w + I = w' + I$ and $z + I = z' + I$. Then $w - w' \in I$ and $z - z' \in I$. By bilinearity of the Lie bracket in L,

$$[w', z'] = [w' + (w - w'), z' + (z - z')]$$
$$= [w, z] + [w - w', z'] + [w', z - z'] + [w - w', z - z'],$$

where the final three summands all belong to I. Therefore $[w' + I, z' + I] = [w, z] + I$, as we needed. It now follows from part (i) of the exercise below that L/I is a Lie algebra. It is called the *quotient* or *factor algebra* of L by I.

Exercise 2.3

(i) Show that the Lie bracket defined on L/I is bilinear and satisfies the axioms (L1) and (L2).

(ii) Show that the linear transformation $\pi : L \to L/I$ which takes an element $z \in L$ to its coset $z + I$ is a homomorphism of Lie algebras.

The reader will not be surprised to learn that there are isomorphism theorems for Lie algebras just as there are for vector spaces and for groups.

Theorem 2.2 (Isomorphism theorems)

(a) Let $\varphi : L_1 \to L_2$ be a homomorphism of Lie algebras. Then $\ker \varphi$ is an ideal of L_1 and $\operatorname{im} \varphi$ is a subalgebra of L_2, and

$$L_1 / \ker \varphi \cong \operatorname{im} \varphi.$$

(b) If I and J are ideals of a Lie algebra, then $(I + J)/J \cong I/(I \cap J)$.

(c) Suppose that I and J are ideals of a Lie algebra L such that $I \subseteq J$. Then J/I is an ideal of L/I and $(L/I)/(J/I) \cong L/J$.

Proof

That $\ker \varphi$ is an ideal of L_1 and $\operatorname{im} \varphi$ is a subalgebra of L_2 were proved in Exercise 1.6. All the isomorphisms we need are already known for vector spaces and their subspaces (see Appendix A): By part (ii) of Exercise 2.3, they are also homomorphisms of Lie algebras. □

Parts (a), (b), and (c) of this theorem are known respectively as the *first*, *second*, and *third isomorphism theorems*.

Example 2.3

Recall that the trace of an $n \times n$ matrix is the sum of its diagonal entries. Fix a field F and consider the linear map $\mathrm{tr} : \mathsf{gl}(n, F) \to F$ which sends a matrix to its trace. This is a Lie algebra homomorphism, for if $x, y \in \mathsf{gl}(n, F)$ then

$$\mathrm{tr}[x, y] = \mathrm{tr}(xy - yx) = \mathrm{tr}\, xy - \mathrm{tr}\, yx = 0,$$

so $\mathrm{tr}[x, y] = [\mathrm{tr}\, x, \mathrm{tr}\, y] = 0$. Here the first Lie bracket is taken in $\mathsf{gl}(n, F)$ and the second in the abelian Lie algebra F.

It is not hard to see that tr is surjective. Its kernel is $\mathsf{sl}(n, F)$, the Lie algebra of matrices with trace 0. Applying the first isomorphism theorem gives

$$\mathsf{gl}(n, F)/\mathsf{sl}(n, F) \cong F.$$

We can describe the elements of the factor Lie algebra explicitly: The coset $x + \mathsf{sl}_n(F)$ consists of those $n \times n$ matrices whose trace is $\mathrm{tr}\, x$.

Exercise 2.4

Show that if L is a Lie algebra then $L/Z(L)$ is isomorphic to a subalgebra of $\mathsf{gl}(L)$.

2.3 Correspondence between Ideals

Suppose that I is an ideal of the Lie algebra L. There is a bijective correspondence between the ideals of the factor algebra L/I and the ideals of L that contain I. This correspondence is as follows. If J is an ideal of L containing I, then J/I is an ideal of L/I. Conversely, if K is an ideal of L/I, then set $J := \{z \in L : z + I \in K\}$. One can readily check that J is an ideal of L and that J contains I. These two maps are inverses of one another.

Example 2.4

Suppose that L is a Lie algebra and I is an ideal in L such that L/I is abelian. In this case, the ideals of L/I are just the subspaces of L/I. By the ideal correspondence, the ideals of L which contain I are exactly the subspaces of L which contain I.

EXERCISES

2.5.† Show that if $z \in L'$ then tr ad $z = 0$.

2.6. Suppose L_1 and L_2 are Lie algebras. Let $L := \{(x_1, x_2) : x_i \in L_i\}$ be the direct sum of their underlying vector spaces. Show that if we define
$$[(x_1, x_2), (y_1, y_2)] := ([x_1, y_1], [x_2, y_2])$$
then L becomes a Lie algebra, the *direct sum* of L_1 and L_2. As for vector spaces, we denote the direct sum of Lie algebras L_1 and L_2 by $L = L_1 \oplus L_2$.

(i) Prove that $\mathsf{gl}(2, \mathbf{C})$ is isomorphic to the direct sum of $\mathsf{sl}(2, \mathbf{C})$ with $\mathbf{C}$, the 1-dimensional complex abelian Lie algebra.

(ii) Show that if $L = L_1 \oplus L_2$ then $Z(L) = Z(L_1) \oplus Z(L_2)$ and $L' = L_1' \oplus L_2'$. Formulate a general version for a direct sum $L_1 \oplus \ldots \oplus L_k$.

(iii) Are the summands in the direct sum decomposition of a Lie algebra uniquely determined? *Hint*: If you think the answer is yes, now might be a good time to read §16.4 in Appendix A on the "diagonal fallacy". The next question looks at this point in more detail.

2.7. Suppose that $L = L_1 \oplus L_2$ is the direct sum of two Lie algebras.

(i) Show that $\{(x_1, 0) : x_1 \in L_1\}$ is an ideal of L isomorphic to L_1 and that $\{(0, x_2) : x_2 \in L_2\}$ is an ideal of L isomorphic to L_2. Show that the projections $p_1(x_1, x_2) = x_1$ and $p_2(x_1, x_2) = x_2$ are Lie algebra homomorphisms.

Now suppose that L_1 and L_2 do not have any non-trivial proper ideals.

(ii) Let J be a proper ideal of L. Show that if $J \cap L_1 = 0$ and $J \cap L_2 = 0$, then the projections $p_1 : J \to L_1$ and $p_2 : J \to L_2$ are isomorphisms.

(iii) Deduce that if L_1 and L_2 are not isomorphic as Lie algebras, then $L_1 \oplus L_2$ has only two non-trivial proper ideals.

(iv) Assume that the ground field is infinite. Show that if $L_1 \cong L_2$ and L_1 is 1-dimensional, then $L_1 \oplus L_2$ has infinitely many different ideals.

2.8. Let L_1 and L_2 be Lie algebras, and let $\varphi : L_1 \to L_2$ be a surjective Lie algebra homomorphism. True or false:

(a)† $\varphi(L_1') = L_2'$;

(b) $\varphi(Z(L_1)) = Z(L_2)$;

(c) if $h \in L_1$ and $\operatorname{ad} h$ is diagonalisable then $\operatorname{ad} \varphi(h)$ is diagonalisable.

What is different if φ is an isomorphism?

2.9.   For each pair of the following Lie algebras over $\mathbf{R}$, decide whether or not they are isomorphic:

   (i) the Lie algebra $\mathbf{R}_\wedge^3$ where the Lie bracket is given by the vector product;

   (ii) the upper triangular 2×2 matrices over $\mathbf{R}$;

   (iii) the strictly upper triangular 3×3 matrices over $\mathbf{R}$;

   (iv) $L = \{x \in \mathsf{gl}(3, \mathbf{R}) : x^t = -x\}$.

   Hint: Use Exercises 1.15 and 2.8.

2.10.   Let F be a field. Show that the derived algebra of $\mathsf{gl}(n, F)$ is $\mathsf{sl}(n, F)$.

2.11.†   In Exercise 1.15, we defined the Lie algebra $\mathsf{gl}_S(n, F)$ over a field F where S is an $n \times n$ matrix with entries in F.

   Suppose that $T \in \mathsf{gl}(n, F)$ is another $n \times n$ matrix such that $T = P^t S P$ for some invertible $n \times n$ matrix $P \in \mathsf{gl}(n, F)$. (Equivalently, the bilinear forms defined by S and T are congruent.) Show that the Lie algebras $\mathsf{gl}_S(n, F)$ and $\mathsf{gl}_T(n, F)$ are isomorphic.

2.12.   Let S be an $n \times n$ invertible matrix with entries in $\mathbf{C}$. Show that if $x \in \mathsf{gl}_S(n, \mathbf{C})$, then $\operatorname{tr} x = 0$.

2.13.   Let I be an ideal of a Lie algebra L. Let B be the centraliser of I in L; that is,

$$B = C_L(I) = \{x \in L : [x, a] = 0 \text{ for all } a \in I\}.$$

Show that B is an ideal of L. Now suppose that

(1) $Z(I) = 0$, and

(2) if $D : I \to I$ is a derivation, then $D = \operatorname{ad} x$ for some $x \in I$.

Show that $L = I \oplus B$.

2.14.†*   Recall that if L is a Lie algebra, we defined L' to be the subspace spanned by the commutators $[x, y]$ for $x, y \in L$. The purpose of this exercise, which may safely be skipped on first reading, is to show that the *set* of commutators may not even be a vector space (and so certainly not an ideal of L).

Let $\mathbf{R}[x, y]$ denote the ring of all real polynomials in two variables. Let L be the set of all matrices of the form

$$A(f(x), g(y), h(x, y)) = \begin{pmatrix} 0 & f(x) & h(x, y) \\ 0 & 0 & g(y) \\ 0 & 0 & 0 \end{pmatrix}.$$

(i) Prove that L is a Lie algebra with the usual commutator bracket. (In contrast to all the Lie algebras seen so far, L is infinite-dimensional.)

(ii) Prove that

$$[A(f_1(x), g_1(y), h_1(x, y)), A(f_2(x), g_2(y), h_2(x, y))] =$$
$$A(0, 0, f_1(x)g_2(y) - f_2(x)g_1(y)).$$

Hence describe L'.

(iii) Show that if $h(x, y) = x^2 + xy + y^2$, then $A(0, 0, h(x, y))$ is not a commutator.

3
Low-Dimensional Lie Algebras

We would like to know how many essentially different (that is, non-isomorphic) Lie algebras there are and what approaches we can use to classify them. To get some feeling for these questions, we shall look at Lie algebras of dimensions 1, 2, and 3. Another reason for looking at these low-dimensional Lie algebras is that they often occur as subalgebras of the larger Lie algebras we shall meet later.

Abelian Lie algebras are easily understood: For any natural number n, there is an abelian Lie algebra of dimension n (where for any two elements, the Lie bracket is zero). We saw in Exercise 1.11 that any two abelian Lie algebras of the same dimension over the same field are isomorphic, so we understand them completely, and from now on we shall only consider non-abelian Lie algebras.

How can we get going? We know that Lie algebras of different dimensions cannot be isomorphic. Moreover, if L is a non-abelian Lie algebra, then its derived algebra L' is non-zero and its centre $Z(L)$ is a proper ideal. By Exercise 2.8, derived algebras and centres are preserved under isomorphism, so it seems reasonable to use the dimension of L and properties of L' and $Z(L)$ as criteria to organise our search.

3.1 Dimensions 1 and 2

Any 1-dimensional Lie algebra is abelian.

Suppose L is a non-abelian Lie algebra of dimension 2 over a field F. The derived algebra of L cannot be more than 1-dimensional since if $\{x, y\}$ is a basis of L, then L' is spanned by $[x, y]$. On the other hand, the derived algebra must be non-zero, as otherwise L would be abelian.

Therefore L' must be 1-dimensional. Take a non-zero element $x \in L'$ and extend it in any way to a vector space basis $\{x, \tilde{y}\}$ of L. Then $[x, \tilde{y}] \in L'$: This element must be non-zero, as otherwise L would be abelian. So there is a non-zero scalar $\alpha \in F$ such that $[x, \tilde{y}] = \alpha x$. This scalar factor does not contribute anything to the structure of L, for if we replace $\tilde{y}$ with $y := \alpha^{-1}\tilde{y}$, then we get

$$[x, y] = x.$$

We have shown that if a 2-dimensional non-abelian Lie algebra exists, then it must have a basis $\{x, y\}$ with the Lie bracket given by the equation above. We should also check that defining the Lie bracket in this way really does give a Lie algebra. In this case, this is straightforward (see Exercise 3.4 for one reason why the Jacobi identity must hold) so we have proved the following theorem.

Theorem 3.1

Let F be any field. Up to isomorphism there is a unique two-dimensional non-abelian Lie algebra over F. This Lie algebra has a basis $\{x, y\}$ such that its Lie bracket is described by $[x, y] = x$. The centre of this Lie algebra is 0. $\square$

When we say "the Lie bracket is described by . . . ," this implicitly includes the information that $[x, x] = 0$ and $[x, y] = -[y, x]$.

3.2 Dimension 3

If L is a non-abelian 3-dimensional Lie algebra over a field F, then we know only that the derived algebra L' is non-zero. It might have dimension 1 or 2 or even 3. We also know that the centre $Z(L)$ is a proper ideal of L. We organise our search by relating L' to $Z(L)$.

3.2.1 The Heisenberg Algebra

Assume first that L' is 1-dimensional and that L' is contained in $Z(L)$. We shall show that there is a unique such Lie algebra, and that it has a basis f, g, z, where $[f, g] = z$ and z lies in $Z(L)$. This Lie algebra is known as the *Heisenberg algebra*.

Take any $f, g \in L$ such that $[f, g]$ is non-zero; as we have assumed that L' is 1-dimensional, the commutator $[f, g]$ spans L'. We have also assumed that L' is contained in the centre of L, so we know that $[f, g]$ commutes with all elements of L. Now set

$$z := [f, g].$$

We leave it as an exercise for the reader to check that f, g, z are linearly independent and therefore form a basis of L. As before, all other Lie brackets are already fixed. In this case, to confirm that this really defines a Lie algebra, we observe that the Lie algebra of strictly upper triangular 3×3 matrices over F has this form if one takes the basis

$$\{e_{12},\ e_{23},\ e_{13}\}.$$

Moreover, we see that L' is in fact equal to the centre $Z(L)$.

3.2.2 Another Lie Algebra where $\dim L' = 1$

The remaining case occurs when L' is 1-dimensional and L' is not contained in the centre of L. We can use the direct sum construction introduced in Exercise 2.6 to give one such Lie algebra. Namely, take $L = L_1 \oplus L_2$, where L_1 is 2-dimensional and non-abelian (that is, the algebra which we found in §3.1) and L_2 is 1-dimensional. By Exercise 2.6,

$$L' = L_1' \oplus L_2' = L_1'$$

and hence L' is 1-dimensional. Moreover, $Z(L) = Z(L_1) \oplus Z(L_2) = L_2$ and therefore L' is not contained in L_2.

Perhaps surprisingly, there are no other Lie algebras with this property. We shall now prove the following theorem.

Theorem 3.2

Let F be any field. There is a unique 3-dimensional Lie algebra over F such that L' is 1-dimensional and L' is not contained in $Z(L)$. This Lie algebra is the direct sum of the 2-dimensional non-abelian Lie algebra with the 1-dimensional Lie algebra.

Proof

We start by picking some non-zero element $x \in L'$. Since x is not central, there must be some $y \in L$ with $[x, y] \neq 0$. By Exercise 1.1, x, y are linearly independent. Since x spans L', we know that $[x, y]$ is a multiple of x. By replacing y with a scalar multiple of itself, we may arrange that $[x, y] = x$. (Alternatively, we could argue that the subalgebra of L generated by x, y is a 2-dimensional non-abelian Lie algebra, so by Theorem 3.2 we may assume that $[x, y] = x$.)

We extend $\{x, y\}$ to a basis of L, say by w. Since x spans L', there exist scalars α, β such that

$$[x, w] = \alpha x, \quad [y, w] = \beta x.$$

We claim that L contains a non-zero central element z which is not in the span of x and y.

For $z = \lambda x + \mu y + \nu w \in L$, we calculate that

$$[x, z] = [x, \lambda x + \mu y + \nu w] = \mu x + \nu \alpha x,$$
$$[y, z] = [y, \lambda x + \mu y + \nu w] = -\lambda x + \nu \beta x.$$

Hence, if we take $\lambda = \beta$, $\mu = -\alpha$, and $\nu = 1$ then $[x, z] = [y, z] = 0$ and z is not in the space spanned by x and y. Hence $L = \mathrm{Span}\,\{x, y\} \oplus \mathrm{Span}\,\{z\}$ is a direct sum of Lie algebras of the required form. $\square$

3.2.3 Lie Algebras with a 2-Dimensional Derived Algebra

Suppose that $\dim L = 3$ and $\dim L' = 2$. We shall see that, over $\mathbf{C}$ at least, there are infinitely many non-isomorphic such Lie algebras.

Take a basis of L', say $\{y, z\}$, and extend it to a basis of L, say by x. To understand the Lie algebra L, we need to understand the structure of L' as a Lie algebra in its own right and how the linear map $\mathrm{ad}\, x : L \to L$ acts on L'. Luckily, this is not too difficult.

Lemma 3.3

(a) The derived algebra L' is abelian.

(b) The linear map $\mathrm{ad}\, x : L' \to L'$ is an isomorphism.

Proof

For part (a), it suffices to show that $[y, z] = 0$. We know that $[y, z]$ lies in L', so there are scalars α and β such that

$$[y, z] = \alpha y + \beta z.$$

Write the matrix of $\operatorname{ad} y$ with respect to the basis x, y, z. It has the form

$$\begin{pmatrix} 0 & 0 & 0 \\ \star & 0 & \alpha \\ \star & 0 & \beta \end{pmatrix},$$

where $\star$ denotes a coefficient we have no need to name explicitly. We see that $\operatorname{tr}(\operatorname{ad} y) = \beta$. As $y \in L'$, Exercise 2.5 implies that $\beta = 0$. Similarly, by considering a matrix for $\operatorname{ad} z$, we get $\alpha = 0$. This proves that $[y, z] = 0$.

Now for part (b). The derived algebra L' is spanned by $[x, y]$, $[x, z]$, and $[y, z]$. However $[y, z] = 0$, so as L' is 2-dimensional, we deduce that $\{[x, y], [x, z]\}$ is a basis of L'. Thus the image of $\operatorname{ad} x$ is 2-dimensional, and $\operatorname{ad} x : L' \to L'$ is an isomorphism. $\qquad\qquad\square$

We shall now try to classify the complex Lie algebras of this form.

Case 1: There is some $x \notin L'$ such that $\operatorname{ad} x : L' \to L'$ is diagonalisable. In this case, we may assume that y, z are eigenvectors of $\operatorname{ad} x$; the associated eigenvalues must be non-zero by part (b) of the lemma.

Suppose that $[x, y] = \lambda y$. We may assume that $\lambda = 1$, for if we scale x by λ^{-1}, we have $[\lambda^{-1} x, y] = y$. With respect to the basis $\{y, z\}$ of L', the linear map $\operatorname{ad} x : L' \to L'$ has matrix

$$\begin{pmatrix} 1 & 0 \\ 0 & \mu \end{pmatrix}$$

for some non-zero $\mu \in \mathbf{C}$.

In Exercise 3.1 below, you are asked to check that these data do define a Lie algebra having the properties with which we started. Call this algebra L_μ. Now we have to decide when two such Lie algebras are isomorphic. In Exercise 3.2, you are asked to prove that L_μ is isomorphic to L_ν if and only if either $\mu = \nu$ or $\mu = \nu^{-1}$. For a solution, see Appendix E. Thus there is an infinite family of non-isomorphic such Lie algebras.

Case 2: For all $x \notin L'$, the linear map $\operatorname{ad} x$ is not diagonalisable. Take any $x \notin L'$. As we work over $\mathbf{C}$, $\operatorname{ad} x : L' \to L'$ must have an eigenvector, say $y \in L'$. As before, by scaling x we may assume that $[x, y] = y$. Extend y

to a basis $\{y, z\}$ of L'. We have $[x, z] = \lambda y + \mu z$ where $\lambda \neq 0$ (otherwise $\operatorname{ad} x$ would be diagonalisable). By scaling z, we may arrange that $\lambda = 1$.

The matrix of $\operatorname{ad} x$ acting on L' therefore has the form:

$$A = \begin{pmatrix} 1 & 1 \\ 0 & \mu \end{pmatrix}.$$

We assumed that A is not diagonalisable, and therefore it cannot have two distinct eigenvalues. It follows that $\mu = 1$.

Again this completely determines a Lie algebra having the properties with which we started. Up to isomorphism, we get just one such algebra.

3.2.4 Lie Algebras where $L' = L$

Suppose that L is a complex Lie algebra of dimension 3 such that $L = L'$. We already know one example, namely $L = \mathsf{sl}(2, \mathbf{C})$. We shall show that up to isomorphism it is the only one.

Step 1: Let $x \in L$ be non-zero. We claim that $\operatorname{ad} x$ has rank 2. Extend x to a basis of L, say $\{x, y, z\}$. Then L' is spanned by $\{[x, y], [x, z], [y, z]\}$. But $L' = L$, so this set must be linearly independent, and hence the image of $\operatorname{ad} x$ has a basis $\{[x, y], [x, z]\}$ of size 2, as required.

Step 2: We claim that there is some $h \in L$ such that $\operatorname{ad} h : L \to L$ has an eigenvector with a non-zero eigenvalue. Choose any non-zero $x \in L$. If $\operatorname{ad} x$ has a non-zero eigenvalue, then we may take $h = x$. If $\operatorname{ad} x : L \to L$ has no non-zero eigenvalues, then, as it has rank 2, its Jordan canonical form (see Appendix A) must be

$$\begin{pmatrix} 0 & 1 & 0 \\ 0 & 0 & 1 \\ 0 & 0 & 0 \end{pmatrix}.$$

This matrix indicates there is a basis of L extending $\{x\}$, say $\{x, y, z\}$, such that $[x, y] = x$ and $[x, z] = y$. So $\operatorname{ad} y$ has x as an eigenvector with eigenvalue -1, and we may take $h = y$.

Step 3: By the previous step, we may find $h, x \in L$ such that $[h, x] = \alpha x \neq 0$. Since $h \in L$ and $L = L'$, we know from Exercise 2.5 that $\operatorname{ad} h$ has trace zero. It follows that $\operatorname{ad} h$ must have three distinct eigenvalues $\alpha, 0, -\alpha$. If y is an eigenvector for $\operatorname{ad} h$ with eigenvalue $-\alpha$, then $\{h, x, y\}$ is a basis of L. In this basis, $\operatorname{ad} h$ is represented by a diagonal matrix.

Step 4: To fully describe the structure of L, we need to determine $[x, y]$. Note that

$$[h, [x, y]] = [[h, x], y] + [x, [h, y]] = \alpha[x, y] + (-\alpha)[x, y] = 0.$$

We now make two applications of Step 1. Firstly, $\ker \operatorname{ad} h = \operatorname{Span}\{h\}$, so $[x, y] = \lambda h$ for some $\lambda \in \mathbf{C}$. Secondly, $\lambda \neq 0$, as otherwise $\ker \operatorname{ad} x$ is 2-dimensional. By replacing x with $\lambda^{-1}x$ we may assume that $\lambda = 1$.

How many such non-isomorphic algebras are there? If in Step 3 we replace h by a non-zero multiple of itself, then we can get any non-zero value for α that we like. In particular, we may take $\alpha = 2$, in which case the structure constants of L with respect to the basis $\{x, y, h\}$ will agree with the structure constants of $\mathsf{sl}(2, \mathbf{C})$ found in Exercise 1.12. Therefore $L \cong \mathsf{sl}(2, \mathbf{C})$. This shows that there is one and only one 3-dimensional complex Lie algebra with $L' = L$.

EXERCISES

3.1. Let V be a vector space and let φ be an endomorphism of V. Let L have underlying vector space $V \oplus \operatorname{Span}\{x\}$. Show that if we define the Lie bracket on L by $[y, z] = 0$ and $[x, y] = \varphi(y)$ for $y, z \in V$, then L is a Lie algebra and $\dim L' = \operatorname{rank} \varphi$. (For a more general construction, see Exercise 3.9 below.)

3.2.† With the notation of §3.2.3, show that the Lie algebra L_μ is isomorphic to L_ν if and only if either $\mu = \nu$ or $\mu = \nu^{-1}$.

3.3. Find out where each of the following 3-dimensional complex Lie algebras appears in our classification:

(i) $\mathsf{gl}_S(3, \mathbf{C})$, where $S = \begin{pmatrix} 1 & 0 & 0 \\ 0 & 1 & 0 \\ 0 & 0 & -1 \end{pmatrix}$;

(ii) the Lie subalgebra of $\mathsf{gl}(3, \mathbf{C})$ spanned by the matrices

$$u = \begin{pmatrix} \lambda & 0 & 0 \\ 0 & \mu & 0 \\ 0 & 0 & \nu \end{pmatrix}, \quad v = \begin{pmatrix} 0 & 0 & 1 \\ 0 & 0 & 0 \\ 0 & 0 & 0 \end{pmatrix}, \quad w = \begin{pmatrix} 0 & 0 & 0 \\ 0 & 0 & 1 \\ 0 & 0 & 0 \end{pmatrix},$$

where λ, μ, ν are fixed complex numbers;

(iii) $\left\{ \begin{pmatrix} 0 & a & b & 0 \\ 0 & 0 & c & 0 \\ 0 & 0 & 0 & 0 \\ 0 & 0 & 0 & 0 \end{pmatrix} : a, b, c \in \mathbf{C} \right\}$;

(iv) $\left\{ \begin{pmatrix} 0 & 0 & a & b \\ 0 & 0 & 0 & c \\ 0 & 0 & 0 & 0 \\ 0 & 0 & 0 & 0 \end{pmatrix} : a, b, c \in \mathbf{C} \right\}.$

3.4.   Suppose L is a vector space with basis x, y and that a bilinear operation $[-, -]$ on L is defined such that $[u, u] = 0$ for all $u \in L$. Show that the Jacobi identity holds and hence L is a Lie algebra.

3.5.   Show that over $\mathbf{R}$ the Lie algebras $\mathsf{sl}(2, \mathbf{R})$ and $\mathbf{R}^3_\wedge$ are not isomorphic. *Hint*: Prove that there is no non-zero $x \in \mathbf{R}^3_\wedge$ such that the map $\operatorname{ad} x$ is diagonalisable.

3.6.†   Show that over $\mathbf{R}$ there are exactly two non-isomorphic 3-dimensional Lie algebras with $L' = L$.

3.7.   Let L be a non-abelian Lie algebra. Show that $\dim Z(L) \leq \dim L - 2$.

3.8.*   Let L be the 3-dimensional Heisenberg Lie algebra defined over a field F. Show that $\operatorname{Der} L$ is 6-dimensional. Identify the inner derivations (as defined in Exercise 1.18) and show that the quotient $\operatorname{Der} L / \operatorname{IDer} L$ is isomorphic to $\mathsf{gl}(2, F)$.

3.9.   Suppose that I is an ideal of a Lie algebra L and that there is a subalgebra S of L such that $L = S \oplus I$.

(i) Show that the map $\theta : S \to \mathsf{gl}(I)$ defined by $\theta(s)x = [s, x]$ is a Lie algebra homomorphism from S into $\operatorname{Der} I$.

We say that L is a *semidirect product* of I by S. (The reader may have seen the analogous construction for groups.)

(ii) Show conversely that given Lie algebras S and I and a Lie algebra homomorphism $\theta : S \to \operatorname{Der} I$, the vector space $S \oplus I$ may be made into a Lie algebra by defining

$$[(s_1, x_1), (s_2, x_2)] = ([s_1, s_2], [x_1, x_2] + \theta(s_1)x_2 - \theta(s_2)x_1)$$

for $s_1, s_2 \in S$, and $x_1, x_2 \in I$, and that this Lie algebra is a semidirect product of I by S. (The direct sum construction introduced in Exercise 2.6 is the special case where $\theta(s) = 0$ for all $s \in S$.)

(iii) Show that the Lie algebras in Exercise 3.1 may be constructed as semidirect products.

(iv)* Investigate necessary and sufficient conditions for two semidirect products to be isomorphic.

3.10.*   Find, up to isomorphism, all Lie algebras with a 1-dimensional derived algebra.

4
Solvable Lie Algebras and a Rough Classification

Abelian Lie algebras are easily understood. There is a sense in which some of the low-dimensional Lie algebras we studied in Chapter 3 are close to being abelian. For example, the 3-dimensional Heisenberg algebra discussed in §3.2.1 has a 1-dimensional centre. The quotient algebra modulo this ideal is also abelian. We ask when something similar might hold more generally. That is, to what extent can we "approximate" a Lie algebra by abelian Lie algebras?

4.1 Solvable Lie Algebras

To start, we take an ideal I of a Lie algebra L and ask when the factor algebra L/I is abelian. The following lemma provides the answer.

Lemma 4.1

Suppose that I is an ideal of L. Then L/I is abelian if and only if I contains the derived algebra L'.

Proof

The algebra L/I is abelian if and only if for all $x, y \in L$ we have

$$[x + I, y + I] = [x, y] + I = I$$

or, equivalently, for all $x, y \in L$ we have $[x, y] \in I$. Since I is a subspace of L, this holds if and only if the space spanned by the brackets $[x, y]$ is contained in I; that is, $L' \subseteq I$.   □

   This lemma tells us that the derived algebra L' is the smallest ideal of L with an abelian quotient. By the same argument, the derived algebra L' itself has a smallest ideal whose quotient is abelian, namely the derived algebra of L', which we denote $L^{(2)}$, and so on. We define the *derived series* of L to be the series with terms

$$L^{(1)} = L' \quad \text{and} \quad L^{(k)} = [L^{(k-1)}, L^{(k-1)}] \text{ for } k \geq 2.$$

Then $L \supseteq L^{(1)} \supseteq L^{(2)} \supseteq \ldots$.
   As the product of ideals is an ideal, $L^{(k)}$ is an ideal of L (and not just an ideal of $L^{(k-1)}$).

Definition 4.2

The Lie algebra L is said to be *solvable* if for some $m \geq 1$ we have $L^{(m)} = 0$.

   The Heisenberg algebra is solvable. Similarly, the algebra of upper triangular matrices is solvable (see Exercise 4.5 below). Furthermore, the classification of 2-dimensional Lie algebras in §3.1 shows that any 2-dimensional Lie algebra is solvable. On the other hand, if $L = \mathsf{sl}(2, \mathbf{C})$, then we have seen in Exercise 2.2 that $L = L'$ and therefore $L^{(m)} = L$ for all $m \geq 1$, so $\mathsf{sl}(2, \mathbf{C})$ is not solvable.
   If L is solvable, then the derived series of L provides us with an "approximation" of L by a finite series of ideals with abelian quotients. This also works the other way around.

Lemma 4.3

If L is a Lie algebra with ideals

$$L = I_0 \supseteq I_1 \supseteq \ldots \supseteq I_{m-1} \supseteq I_m = 0$$

such that I_{k-1}/I_k is abelian for $1 \leq k \leq m$, then L is solvable.

Proof

We shall show that $L^{(k)}$ is contained in I_k for k between 1 and m. Putting $k = m$ will then give $L^{(m)} = 0$.

Since L/I_1 is abelian, we have from Lemma 4.1 that $L' \subseteq I_1$. For the inductive step, we suppose that $L^{(k-1)} \subseteq I_{k-1}$, where $k \geq 2$. The Lie algebra I_{k-1}/I_k is abelian. Therefore by Lemma 4.1, this time applied to the Lie algebra I_{k-1}, we have $[I_{k-1}, I_{k-1}] \subseteq I_k$. But $L^{(k-1)}$ is contained in I_{k-1} by our inductive hypothesis, so we deduce that

$$L^{(k)} = [L^{(k-1)}, L^{(k-1)}] \subseteq [I_{k-1}, I_{k-1}],$$

and hence $L^{(k)} \subseteq I_k$. $\qquad\qquad\qquad\square$

This proof shows that if $L^{(k)}$ is non-zero then I_k is also non-zero. Hence the derived series may be thought of as the fastest descending series whose successive quotients are abelian.

Lie algebra homomorphisms are linear maps that preserve Lie brackets, and so one would expect that they preserve the derived series.

Exercise 4.1

Suppose that $\varphi : L_1 \to L_2$ is a surjective homomorphism of Lie algebras. Show that

$$\varphi(L_1^{(k)}) = (L_2)^{(k)}.$$

This exercise suggests that the property of being solvable should be inherited by various constructions.

Lemma 4.4

Let L be a Lie algebra.

(a) If L is solvable, then every subalgebra and every homomorphic image of L are solvable.

(b) Suppose that L has an ideal I such that I and L/I are solvable. Then L is solvable.

(c) If I and J are solvable ideals of L, then $I + J$ is a solvable ideal of L.

Proof

(a) If L_1 is a subalgebra of L, then for each k it is clear that $L_1^{(k)} \subseteq L^{(k)}$, so if $L^{(m)} = 0$, then also $L_1^{(m)} = 0$. For the second part, apply Exercise 4.1.

(b) We have $(L/I)^{(k)} = (L^{(k)} + I)/I$. (Either apply Exercise 4.1 to the canonical homomorphism $L \to L/I$ or prove this directly by induction on k.) If L/I is solvable then for some $m \geq 1$ we have $(L/I)^{(m)} = 0$; that is, $L^{(m)} + I = I$ and therefore $L^{(m)} \subseteq I$. If I is also solvable, then $I^{(k)} = 0$ for some $k \geq 1$ and hence $(L^{(m)})^{(k)} \subseteq I^{(k)} = 0$. Now one can convince oneself that by definition

$$(L^{(m)})^{(k)} = L^{(m+k)}.$$

(c) By the second isomorphism theorem $(I + J)/I \cong J/I \cap J$, so it is solvable by Lemma 4.4(a). Since I is also solvable, part (b) of this lemma implies that $I + J$ is solvable.   □

Corollary 4.5

Let L be a finite-dimensional Lie algebra. There is a unique solvable ideal of L containing every solvable ideal of L.

Proof

Let R be a solvable ideal of largest possible dimension. Suppose that I is any solvable ideal. By Lemma 4.4(c), we know that $R + I$ is a solvable ideal. Now $R \subseteq R + I$ and hence $\dim R \leq \dim(R + I)$. We chose R of maximal possible dimension and therefore we must have $\dim R = \dim(R+I)$ and hence $R = R+I$, so I is contained in R.   □

This largest solvable ideal is said to be the *radical* of L and is denoted $\operatorname{rad} L$. The radical will turn out to be an essential tool in helping to describe the finite-dimensional Lie algebras. It also suggests the following definition.

Definition 4.6

A non-zero finite-dimensional Lie algebra L is said to be *semisimple* if it has no non-zero solvable ideals or equivalently if $\operatorname{rad} L = 0$.

For example, by Exercise 1.13, $\mathsf{sl}(2, \mathbf{C})$ is semisimple. The reason for the word "semisimple" is revealed in §4.3 below.

Lemma 4.7

If L is a Lie algebra, then the factor algebra $L/\operatorname{rad} L$ is semisimple.

Proof

Let $\bar{J}$ be a solvable ideal of $L/\operatorname{rad} L$. By the ideal correspondence, there is an ideal J of L containing $\operatorname{rad} L$ such that $\bar{J} = J/\operatorname{rad} L$. By definition, $\operatorname{rad} L$ is solvable, and $J/\operatorname{rad} L = \bar{J}$ is solvable by hypothesis. Therefore Lemma 4.4 implies that J is solvable. But then J is contained in $\operatorname{rad} L$; that is, $\bar{J} = 0$. $\square$

4.2 Nilpotent Lie Algebras

We define the *lower central series* of a Lie algebra L to be the series with terms

$$L^1 = L' \quad \text{and} \quad L^k = [L, L^{k-1}] \text{ for } k \geq 2.$$

Then $L \supseteq L^1 \supseteq L^2 \supseteq \ldots$. As the product of ideals is an ideal, L^k is even an ideal of L (and not just an ideal of L^{k-1}). The reason for the name "central series" comes from the fact that L^k/L^{k+1} is contained in the centre of L/L^{k+1}.

Definition 4.8

The Lie algebra L is said to be *nilpotent* if for some $m \geq 1$ we have $L^m = 0$.

The Lie algebra $\mathsf{n}(n, F)$ of strict upper triangular matrices over a field F is nilpotent (see Exercise 4.4). Furthermore, any nilpotent Lie algebra is solvable. To see this, show by induction on k that $L^{(k)} \subseteq L^k$. There are solvable Lie algebras which are not nilpotent; the standard example is the Lie algebra $\mathsf{b}(n, F)$ of upper triangular matrices over a field F for $n \geq 2$ (see Exercise 4.5). Another is the two-dimensional non-abelian Lie algebra (see §3.1).

Lemma 4.9

Let L be a Lie algebra.

(a) If L is nilpotent, then any Lie subalgebra of L is nilpotent.

(b) If $L/Z(L)$ is nilpotent, then L is nilpotent.

Proof

Part (a) is clear from the definition. By induction, or by a variation of Exercise 4.1, one can show that $(L/Z(L))^k$ is equal to $\left(L^k + Z(L)\right)/Z(L)$. So if $(L/Z(L))^m$ is zero, then L^m is contained in $Z(L)$ and therefore $L^{m+1} = 0$. $\square$

Remark 4.10

The analogue of Lemma 4.4(b) does not hold; that is, if I is an ideal of a Lie algebra L, then it is possible that both L/I and I are nilpotent but L is not. An example is given by the 2-dimensional non-abelian Lie algebra. This perhaps suggests that solvability is more fundamental to the structure of Lie algebras than nilpotency.

4.3 A Look Ahead

The previous section suggests that we might have a chance to understand all finite-dimensional Lie algebras. The radical rad L of any Lie algebra L is solvable, and $L/\mathrm{rad}\,L$ is semisimple, so to understand L it is necessary to understand

(i) an arbitrary solvable Lie algebra and

(ii) an arbitrary semisimple Lie algebra.

Working over $\mathbf{C}$, an answer to (i) was found by Lie, who proved (in essence) that every solvable Lie algebra appears as a subalgebra of a Lie algebras of upper triangular matrices. We give a precise statement of Lie's Theorem in §6.4 below.

For (ii) we shall show that every semisimple Lie algebra is a direct sum of *simple* Lie algebras.

Definition 4.11

The Lie algebra L is *simple* if it has no ideals other than 0 and L and it is not abelian.

The restriction that a simple Lie algebra may not be abelian removes only the 1-dimensional abelian Lie algebra. Without this restriction, this Lie algebra would be simple but not semisimple: This is obviously undesirable.

We then need to find all simple Lie algebras over $\mathbf{C}$. This is a major theorem; the proof is based on work by Killing, Engel, and Cartan. We shall eventually prove most of the following theorem.

Theorem 4.12 (Simple Lie algebras)

With five exceptions, every finite-dimensional simple Lie algebra over $\mathbf{C}$ is isomorphic to one of the *classical Lie algebras*:

$$\mathsf{sl}(n, \mathbf{C}), \quad \mathsf{so}(n, \mathbf{C}), \quad \mathsf{sp}(2n, \mathbf{C}).$$

The five exceptional Lie algebras are known as e_6, e_7, e_8, f_4, and g_2.

We have already introduced the family of special linear Lie algebras, $\mathsf{sl}(n, \mathbf{C})$. The remaining families can be defined as certain subalgebras of $\mathsf{gl}(n, \mathbf{C})$ using the construction introduced in Exercise 1.15. Recall that if $S \in \mathsf{gl}(n, \mathbf{C})$, then we defined a Lie subalgebra of $\mathsf{gl}(n, \mathbf{C})$ by

$$\mathsf{gl}_S(n, \mathbf{C}) := \left\{ x \in \mathsf{gl}(n, \mathbf{C}) : x^t S = -Sx \right\}.$$

Assume first of all that $n = 2\ell$. Take S to be the matrix with $\ell \times \ell$ blocks:

$$S = \begin{pmatrix} 0 & I_\ell \\ I_\ell & 0 \end{pmatrix}.$$

We define $\mathsf{so}(2\ell, \mathbf{C}) = \mathsf{gl}_S(2\ell, \mathbf{C})$. When $n = 2\ell + 1$, we take

$$S = \begin{pmatrix} 1 & 0 & 0 \\ 0 & 0 & I_\ell \\ 0 & I_\ell & 0 \end{pmatrix}$$

and define $\mathsf{so}(2\ell + 1, \mathbf{C}) = \mathsf{gl}_S(2\ell + 1, \mathbf{C})$. These Lie algebras are known as the *orthogonal Lie algebras*.

The Lie algebras $\mathsf{sp}(n, \mathbf{C})$ are only defined for even n. If $n = 2\ell$, we take

$$S = \begin{pmatrix} 0 & I_\ell \\ -I_\ell & 0 \end{pmatrix}$$

and define $\mathsf{sp}(2\ell, \mathbf{C}) = \mathsf{gl}_S(2\ell, \mathbf{C})$. These Lie algebras are known as the *symplectic Lie algebras*.

It follows from Exercise 2.12 that $\mathsf{so}(n, \mathbf{C})$ and $\mathsf{sp}(n, \mathbf{C})$ are subalgebras of $\mathsf{sl}(n, \mathbf{C})$. (This also follows from the explicit bases given in Chapter 12.)

We postpone discussion of the exceptional Lie algebras until Chapter 14.

Exercise 4.2

Let $x \in \mathsf{gl}(2\ell, \mathbf{C})$. Show that x belongs to $\mathsf{sp}(2\ell, \mathbf{C})$ if and only if it is of the form

$$x = \begin{pmatrix} m & p \\ q & -m^t \end{pmatrix},$$

where p and q are symmetric. Hence find the dimension of $\mathsf{sp}(2\ell, \mathbf{C})$. (See Exercise 12.1 for the other families.)

EXERCISES

4.3.   Use Lemma 4.4 to show that if L is a Lie algebra then L is solvable
        if and only if $\operatorname{ad} L$ is a solvable subalgebra of $\mathsf{gl}(L)$. Show that this
        result also holds if we replace "solvable" with "nilpotent."

4.4.   Let $L = \mathsf{n}(n, F)$, the Lie algebra of strictly upper triangular $n \times n$
        matrices over a field F. Show that L^k has a basis consisting of all
        the matrix units e_{ij} with $j - i > k$. Hence show that L is nilpotent.
        What is the smallest m such that $L^m = 0$?

4.5.   Let $L = \mathsf{b}(n, F)$ be the Lie algebra of upper triangular $n \times n$ matrices
        over a field F.

   (i) Show that $L' = \mathsf{n}(n, F)$.

   (ii) More generally, show that $L^{(k)}$ has a basis consisting of all the
        matrix units e_{ij} with $j - i \geq 2^{k-1}$. (The commutator formula
        for the e_{ij} given in §1.2 will be helpful.)

   (iii) Hence show that L is solvable. What is the smallest m such
        that $L^{(m)} = 0$?

   (iv) Show that if $n \geq 2$ then L is not nilpotent.

4.6.   Show that a Lie algebra is semisimple if and only if it has no non-
        zero abelian ideals. (This was the original definition of semisimplicity
        given by Wilhelm Killing.)

4.7.   Prove directly that $\mathsf{sl}(n, \mathbf{C})$ is a simple Lie algebra for $n \geq 2$.

4.8.†   Let L be a Lie algebra over a field F such that $[[a, b], b] = 0$ for all
        $a, b \in L$, (or equivalently, $(\operatorname{ad} b)^2 = 0$ for all $b \in L$).

   (i) Suppose the characteristic of F is not 3. Show that then $L^3 = 0$.

   (ii)* Show that if F has characteristic 3 then $L^4 = 0$. *Hint:* show
        first that the Lie brackets $[[x, y], z]$ are alternating; that is,

$$[[x, y], z] = -[[y, x], z], \quad [[x, y], z] = -[[x, z], y]$$

        for all $x, y, z \in L$.

4.9.*   The purpose of this exercise is to give some idea why the families
        of Lie algebras are given the names that we have used. We shall
        not need to refer to this exercise later; some basic group theory is
        needed.

We begin with the Lie algebra $\mathsf{sl}(n, \mathbf{C})$. Recall that the $n \times n$ matrices with determinant 1 form a group under matrix multiplication, denoted $\mathrm{SL}(n, \mathbf{C})$. Let I denote the $n \times n$ identity matrix. We ask: when is $I + \varepsilon X \in \mathrm{SL}(n, \mathbf{C})$ for X an $n \times n$ matrix?

(i) Show that $\det(I + \varepsilon X)$ is a polynomial in ε of degree n with the first two terms

$$\det(I + \varepsilon X) = 1 + (\operatorname{tr} X)\varepsilon + \ldots.$$

If we neglect all powers of ε except 1 and ε, then we obtain the statement

$$I + \varepsilon X \in \mathrm{SL}(n, \mathbf{C}) \iff X \in \mathsf{sl}(n, \mathbf{C}).$$

This could have been taken as the definition of $\mathsf{sl}(n, \mathbf{C})$. (This is despite the fact that, interpreted literally, it is false!)

(ii) (a) Let S be an $n \times n$ matrix. Let $(-, -)$ denote the complex bilinear form with matrix S. Show that if we let $G_S(n, \mathbf{C})$ be the set of invertible matrices A such that $(Av, Av) = (v, v)$ for all $v \in \mathbf{C}^n$, then $G_S(n, \mathbf{C})$ is a group.

(b) Show that if we perform the construction in part (i) with $G_S(n, \mathbf{C})$ in place of $\mathrm{SL}(n, \mathbf{C})$, we obtain $\mathsf{gl}_S(n, \mathbf{C})$.

(iii) (a) An invertible matrix A is customarily said to be *orthogonal* if $A^t A = A A^t = I$; that is, if $A^{-1} = A^t$. Show that the set of $n \times n$ orthogonal matrices with coefficients in $\mathbf{C}$ is the group $G_I(n, \mathbf{C})$ and that the associated Lie algebra, $\mathsf{g}_I(n, \mathbf{C})$, is the space of all anti-symmetric matrices.

(b) Prove that $\mathsf{g}_I(n, \mathbf{C}) \cong \mathsf{so}(n, \mathbf{C})$. *Hint*: Use Exercise 2.11. (The reason for not taking this as the definition of $\mathsf{so}(n, \mathbf{C})$ will emerge.)

(iv) A bilinear form (see Appendix A) on a vector space v is said to be *symplectic* if $(v, v) = 0$ for all $v \in V$. Show that

$$S = \begin{pmatrix} 0 & I_\ell \\ -I_\ell & 0 \end{pmatrix}$$

is the matrix of a non-degenerate symplectic bilinear form on a 2ℓ-dimensional space. The associated Lie algebra is $\mathsf{gl}_S(2\ell, \mathbf{C}) = \mathsf{sp}(2\ell, \mathbf{C})$.

The reader is entitled to feel rather suspicious about our cavalier treatment of the powers of ε. A rigorous and more general treatment is given in books on matrix groups and Lie groups, such as *Matrix Groups* by Baker [3] in the SUMS series. We shall not attempt to go any further in this direction.

4.10.* Let F be a field. Exercise 2.11 shows that if $S, T \in \mathrm{gl}(n, F)$ are congruent matrices (that is, there exists an invertible matrix P such that $T = P^t S P$), then $\mathrm{gl}_S(n, F) \cong \mathrm{gl}_T(n, F)$. Does the converse hold when $F = \mathbf{C}$? For a challenge, think about other fields.

Subalgebras of gl(V)

Many Lie algebras occur naturally as subalgebras of the Lie algebras of linear transformations of vector spaces. Even more are easily seen to be isomorphic to such subalgebras. Given such a Lie algebra, one can profitably use linear algebra to study its properties.

Throughout this chapter, we let V denote an n-dimensional vector space over a field F. We consider some elementary facts concerning linear maps and Lie subalgebras of gl(V) which are needed for the theorems to come.

5.1 Nilpotent Maps

Let L be a Lie subalgebra of gl(V). We may regard elements of L as linear transformations of V, so in addition to the Lie bracket we can also exploit compositions xy of linear maps for $x, y \in L$. Care must be taken, however, as in general this composition will not belong to L. Suppose that $x \in L$ is a nilpotent map; that is, $x^r = 0$ for some $r \geq 1$. What does this tell us about x as an element of the Lie algebra?

Lemma 5.1·

Let $x \in L$. If the linear map $x : V \to V$ is nilpotent, then $\operatorname{ad} x : L \to L$ is also nilpotent.

Proof

To see this, take $y \in L$ and expand $(\operatorname{ad} x)^m(y) = [x, [x, \ldots [x, y] \ldots]]$. Every term in the resulting sum is of the form $x^j y x^{m-j}$ for some j between 0 and m. For example, $(\operatorname{ad} x)^2 y = x^2 y - 2xyx + yx^2$. Suppose that $x^r = 0$ and let $m \geq 2r$. Either $j \geq r$, in which case $x^j = 0$, or $m - j \geq r$, in which case $x^{m-j} = 0$. It follows that $(\operatorname{ad} x)^{2r} = 0$.   □

This lemma can be regarded as a companion result to Exercise 1.17, which asked you to show that if $L = \operatorname{gl}(V)$ and $x : V \to V$ is diagonalisable, then $\operatorname{ad} x : L \to L$ is also diagonalisable.

5.2 Weights

In linear algebra, one is often interested in the eigenvalues and eigenvectors of a fixed linear map. We now generalise these notions to families of linear maps. Let A be a subalgebra of $\operatorname{gl}(V)$. It seems reasonable to say that $v \in V$ is an *eigenvector for A* if v is an eigenvector for every element of A; that is, $a(v) \in \operatorname{Span}\{v\}$ for every $a \in A$.

Example 5.2

Let $A = \operatorname{d}(n, F)$, the Lie subalgebra of $\operatorname{gl}(n, F)$ consisting of diagonal matrices. Let $\{e_1, \ldots, e_n\}$ be the standard basis of F^n. Then each e_i is an eigenvector for A.

It is not so obvious how we should generalise eigenvalues. Consider the example above. Let a be the diagonal matrix with entries $(\alpha_1, \ldots, \alpha_n)$. The eigenvalue of a on e_i is α_i, but this varies as a runs through the elements of A, so we must be prepared to let different elements of A act with different eigenvalues. We can specify the eigenvalues of elements of A by giving a function $\lambda : A \to F$. The corresponding eigenspace is then

$$V_\lambda := \{v \in V : a(v) = \lambda(a)v \text{ for all } a \in A\}.$$

Exercise 5.1

(i) Check that the eigenspaces V_λ defined above are vector subspaces of V.

(ii) Using the notation above, define $\varepsilon_i : A \to F$ by $\varepsilon_i(a) = \alpha_i$. Show that $V_{\varepsilon_i} = \mathrm{Span}\{e_i\}$ and that V decomposes as a direct sum of the V_{ε_i} for $1 \le i \le n$.

The reader may already have realised that not every function $A \to F$ can have a non-zero eigenspace. Suppose that V_λ is a non-zero eigenspace for the function $\lambda : A \to F$. Let $v \in V_\lambda$ be non-zero, let $a, b \in A$, and let $\alpha, \beta \in F$. Then

$$(\alpha a + \beta b)v = \alpha(av) + \beta(bv) = \alpha\lambda(a)v + \beta\lambda(b)v = (\alpha\lambda(a) + \beta\lambda(b))v,$$

so the eigenvalue of $\alpha a + \beta b$ on v is $\alpha\lambda(a) + \beta\lambda(b)$. In other words, $\lambda(\alpha a + \beta b) = \alpha\lambda(a) + \beta\lambda(b)$. Thus λ is linear and so $\lambda \in A^\star$, the dual space of linear maps from A to F.

We now introduce the standard terminology.

Definition 5.3

A *weight* for a Lie subalgebra A of $\mathsf{gl}(V)$ is a linear map $\lambda : A \to F$ such that

$$V_\lambda := \{v \in V : a(v) = \lambda(a)v \text{ for all } a \in A\}$$

is a non-zero subspace of V.

The vector space V_λ is the *weight space* associated to the weight λ. Thus V_λ is non-zero if and only if V contains a common eigenvector for the elements of A, with the eigenvalues of elements of A given by the function λ.

Exercise 5.2

Let $A = \mathsf{b}(n, F)$ be the Lie subalgebra of $\mathsf{gl}(n, F)$ consisting of upper triangular matrices. Show that e_1 is an eigenvector for A. Find the corresponding weight and determine its weight space.

5.3 The Invariance Lemma

In linear algebra, one shows that if $a, b : V \to V$ are commuting linear transformations and W is the kernel of a, then W is b-invariant. That is, b maps W into W. The proof is very easy: If $w \in W$, then $a(bw) = b(aw) = 0$ and so $bw \in W$. This result has a generalisation to Lie subalgebras of $\mathsf{gl}(V)$ as follows.

Lemma 5.4

Suppose that A is an ideal of a Lie subalgebra L of gl(V). Let

$$W = \{v \in V : a(v) = 0 \text{ for all } a \in A\}.$$

Then W is an L-invariant subspace of V.

Proof

Take $w \in W$ and $y \in L$. We must show that $a(yw) = 0$ for all $a \in A$. But $ay = ya + [a, y]$, where $[a, y] \in A$ as A is an ideal, so

$$a(yw) = y(aw) + [a, y](w) = 0. \qquad \square$$

The technique used here of replacing ay with $ya + [a, y]$ is frequently useful. We shall have recourse to it many times below.

Exercise 5.3

Show that Lemma 5.4 really does generalise the result mentioned in the first paragraph of this section.

This has dealt with zero eigenvalues. More generally, one can prove that if $a, b : V \to V$ are commuting linear maps, $\lambda \in F$, and V_λ is the λ-eigenspace of a (that is, $V_\lambda = \{v \in V : av = \lambda v\}$), then V_λ is invariant under b.

This fact too has a generalisation to Lie algebras. As before, we shall replace the linear map a by an ideal $A \subseteq$ gl(V). The subspace considered in Lemma 5.4 may be viewed as the 0-weight space for A. In our generalisation, we allow an arbitrary weight.

Lemma 5.5 (Invariance Lemma)

Assume that F has characteristic zero. Let L be a Lie subalgebra of gl(V) and let A be an ideal of L. Let $\lambda : A \to F$ be a weight of A. The associated weight space

$$V_\lambda = \{v \in V : av = \lambda(a)v \text{ for all } a \in A\}$$

is an L-invariant subspace of V.

Proof

We must show that if $y \in L$ and $w \in V_\lambda$, then $y(w)$ is an eigenvector for every element of A, with the eigenvalue of $a \in A$ given by $\lambda(a)$.

For $a \in A$, we have

$$a(yw) = y(aw) + [a,y](w) = \lambda(a)yw + \lambda([a,y])w.$$

Note that $[a,y] \in A$ as A is an ideal. Therefore all we need show is that the eigenvalue of the commutator $[a,y]$ on V_λ is zero.

Consider $U = \operatorname{Span}\{w, y(w), y^2(w), \ldots,\}$. This is a finite-dimensional subspace of V. Let m be the least number such that the vectors $w, y(w), \ldots, y^m(w)$ are linearly dependent. It is a straightforward exercise in linear algebra to show that U is m-dimensional and has as a basis

$$w, y(w), \ldots, y^{m-1}(w).$$

We claim that if $z \in A$, then z maps U into itself. In fact, we shall show that with respect to the basis above, z has an upper triangular matrix with diagonal entries equal to $\lambda(z)$:

$$\begin{pmatrix} \lambda(z) & \star & \cdots & \star \\ 0 & \lambda(z) & \cdots & \star \\ \vdots & \vdots & \ddots & \vdots \\ 0 & 0 & \cdots & \lambda(z) \end{pmatrix}.$$

We work by induction on the number of the column. First of all, $zw = \lambda(z)w$. This gives the first column of the matrix. Next, since $[z,y] \in A$, we have

$$z(yw) = y(zw) + [z,y]w = \lambda(z)y(w) + \lambda([z,y])w$$

giving the second column.

More generally, for column $r + 1$, we have

$$z(y^r(w)) = zy(y^{r-1}w) = (yz + [z,y])y^{r-1}w.$$

By the inductive hypothesis, we can say that

$$z(y^{r-1}w) = \lambda(z)y^{r-1}w + u$$

for some u in the span of $\{y^j w : j < r - 1\}$. Substituting this gives

$$yz(y^{r-1}w) = \lambda(z)y^r w + yu,$$

and yu belongs to the span of the $\{y^j w : j < r\}$. Furthermore, since $[z,y] \in A$, we get by induction that

$$[z,y]y^{r-1}w = v$$

for some v in the span of $\{y^j w : j \le r - 1\}$. Combining the last two results shows that column $r + 1$ is as stated.

Now take $z = [a, y]$. We have just shown that the trace of the matrix of z acting on U is $m\lambda(z)$. On the other hand, by the previous paragraph, U is invariant under the action of $a \in A$, and U is y-invariant by construction. So the trace of z on U is the trace of $ay - ya$, also viewed as a linear transformation of U, and this is obviously 0. Therefore

$$m\lambda([a, y]) = 0.$$

As F has characteristic zero, it follows that $\lambda([a, y]) = 0$. $\square$

Remark 5.6

This proof would be much easier if it were true that the linear maps y^r belonged to the Lie algebra L. Unfortunately, this is not true in general.

5.4 An Application of the Invariance Lemma

The following proposition shows how the Invariance Lemma may be applied in practice.

Proposition 5.7

Let $x, y : V \to V$ be linear maps from a complex vector space V to itself. Suppose that x and y both commute with $[x, y]$. Then $[x, y]$ is a nilpotent map.

Proof

Since we are working over the complex numbers, it is enough to show that if λ is an eigenvalue of the linear map $[x, y]$, then $\lambda = 0$.

Let λ be an eigenvalue. Let $W = \{v \in V : [x, y]v = \lambda v\}$ be the corresponding eigenspace of $[x, y]$; this is a non-zero subspace of V. Let L be the Lie subalgebra of $\mathsf{gl}(V)$ spanned by x, y, and $[x, y]$. As Span $\{[x, y]\}$ is an ideal of L, the Invariance Lemma implies that W is invariant under x and y.

Pick a basis of W and let X and Y be the matrices of x and y with respect to this basis. The commutator $[x, y]$ has matrix $XY - YX$ in its action on W.

But every element of W is an eigenvector of $[x, y]$ with eigenvalue λ, so

$$XY - YX = \begin{pmatrix} \lambda & 0 & \cdots & 0 \\ 0 & \lambda & \cdots & 0 \\ \vdots & \vdots & \ddots & \vdots \\ 0 & 0 & \cdots 0 & \lambda \end{pmatrix}.$$

Now taking traces gives

$$0 = \operatorname{tr}(XY - YX) = \lambda \dim W$$

and so $\lambda = 0$. $\qquad\qquad\qquad\qquad\qquad\qquad\qquad\qquad\qquad\qquad\qquad\square$

For another approach, see Exercise 5.5 below. The proposition is also a corollary of Lie's Theorem: See Exercise 6.6 in the next chapter.

EXERCISES

5.4.   (i) Let L be a Lie subalgebra of $\mathsf{gl}(V)$. Suppose that there is a basis of V such that, with respect to this basis, every $x \in L$ is represented by a strictly upper triangular matrix. Show that L is isomorphic to a Lie subalgebra of $\mathsf{n}(n, F)$ and hence that L is nilpotent.

(ii) Prove a result analogous to (i) in the case where there is a basis of V such that with respect to this basis every x in L is represented by an upper triangular matrix.

5.5.⋆ This exercise gives an alternative approach to Proposition 5.7. Recall that V is a complex vector space and $x, y \in \mathsf{gl}(V)$ are linear maps such that $[x, y]$ commutes with both x and y.

(i) Let $z = [x, y]$. Show that $\operatorname{tr} z^m = 0$ for all $m \geq 1$.

(ii) Let $\lambda_1, \ldots, \lambda_n$ be the eigenvalues of $[x, y]$. Show that $\lambda_i = 0$ for all i, and deduce that $[x, y]$ is nilpotent. *Hint*: Use the Vandermonde determinant.

5.6.   Let L be a Lie algebra and let A be a subalgebra of L. The *normaliser* of A, denoted $N_L(A)$, is defined by

$$N_L(A) = \{x \in L : [x, a] \in A \text{ for all } a \in A\}.$$

(i) Prove that $N_L(A)$ is a subalgebra of L and that $N_L(A)$ contains A. Show moreover that $N_L(A)$ is the largest subalgebra of L in which A is an ideal.

(ii) Let $L = \mathsf{gl}(n, \mathbf{C})$ and let A be the subalgebra of L consisting of all diagonal matrices. Show that $N_L(A) = A$.

Hint: While this can be proved directly, it follows very quickly from the ideas in this chapter. By the Invariance Lemma, any weight space for A is $N_L(A)$-invariant. How does this restrict the possible matrices of elements of $N_L(A)$?

5.7.   Show that if $a, y \in \mathsf{gl}(V)$ are linear maps, then for any $m \geq 1$

$$ay^m = y^m a + \sum_{k=1}^{m} \binom{m}{k} y^{m-k} a_k$$

where $a_1 = [a, y]$, $a_2 = [a_1, y] = [[a, y], y]$, and generally $a_k = [a_{k-1}, y]$. Deduce that

$$\mathrm{ad}\, y^m = \sum_{k=1}^{m} (-1)^{k+1} \binom{m}{k} y^{m-k} (\mathrm{ad}\, y)^k.$$

In this formula, the power y^{m-k} acts on $\mathsf{gl}(V)$ as the linear map which sends $x \in \mathsf{gl}(V)$ to $y^{m-k}x$.

6
Engel's Theorem and Lie's Theorem

A useful result in linear algebra states that if V is a finite-dimensional vector space and $x : V \to V$ is a nilpotent linear map, then there is a basis of V in which x is represented by a strictly upper triangular matrix.

To understand Lie algebras, we need a much more general version of this result. Instead of considering a single linear transformation, we consider a Lie subalgebra L of $\mathbf{gl}(V)$. We would like to know when there is a basis of V in which every element of L is represented by a strictly upper triangular matrix.

As a strictly upper triangular matrix is nilpotent, if such a basis exists then every element of L must be a nilpotent map. Surprisingly, this obvious necessary condition is also sufficient; this result is known as Engel's Theorem.

It is also natural to ask the related question: When is there a basis of V in which every element of L is represented by an upper triangular matrix? If there is such a basis, then L is isomorphic to a subalgebra of a Lie algebra of upper triangular matrices, and so L is solvable. Over $\mathbf{C}$ at least, this necessary condition is also sufficient. We prove this result, Lie's Theorem, in §6.4 below.

6.1 Engel's Theorem

Theorem 6.1 (Engel's Theorem)

Let V be a vector space. Suppose that L is a Lie subalgebra of $\mathsf{gl}(V)$ such that every element of L is a nilpotent linear transformation of V. Then there is a basis of V in which every element of L is represented by a strictly upper triangular matrix.

To prove Engel's Theorem, we adapt the strategy used to prove the analogous result for a single nilpotent linear transformation. The proof of this result is outlined in the following exercise.

Exercise 6.1

Let V be an n-dimensional vector space where $n \geq 1$, and let $x : V \to V$ be a nilpotent linear map.

(i) Show that there is a non-zero vector $v \in V$ such that $xv = 0$.

(ii) Let $U = \mathrm{Span}\,\{v\}$. Show that x induces a nilpotent linear transformation $\bar{x} : V/U \to V/U$. By induction, we know that there is a basis $\{v_1 + U, \ldots, v_{n-1} + U\}$ of V/U in which $\bar{x}$ has a strictly upper triangular matrix. Prove that $\{v, v_1, \ldots, v_{n-1}\}$ is a basis of V and that the matrix of x in this basis is strictly upper triangular.

The crucial step in the proof of Engel's Theorem is the analogue of part (i): We must find a non-zero vector $v \in V$ that lies in the kernel of *every* $x \in L$.

Proposition 6.2

Suppose that L is a Lie subalgebra of $\mathsf{gl}(V)$, where V is non-zero, such that every element of L is a nilpotent linear transformation. Then there is some non-zero $v \in V$ such that $xv = 0$ for all $x \in L$.

Proof

We proceed by induction on $\dim L$. If $\dim L = 1$, then L is spanned by a single nilpotent linear transformation, say z. By part (i) of the previous exercise there is some non-zero $v \in V$ such that $zv = 0$. An arbitrary element of L is a scalar multiple of z, so v lies in the kernel of every element of L. Now suppose that $\dim(L) > 1$.

Step 1: Take a maximal Lie subalgebra A of L. We claim that A is an ideal of L and that $\dim A = \dim L - 1$. Consider the quotient vector space $\bar{L} = L/A$. We define a linear map

$$\varphi : A \to \mathsf{gl}(\bar{L})$$

by letting $\varphi(a)$ act on $\bar{L}$ as

$$\varphi(a)(x + A) = [a, x] + A.$$

This is well-defined, for if $x \in A$ then $[a, x] \in A$. Moreover, φ is a Lie homomorphism, for if $a, b \in A$ then

$$
\begin{aligned}
[\varphi(a), \varphi(b)](x + A) &= \varphi(a)([b, x] + A) - \varphi(b)([a, x] + A) \\
&= ([a, [b, x]] + A) - ([b, [a, x]] + A) \\
&= [a, [b, x]] - [b, [a, x]] + A \\
&= [[a, b], x] + A
\end{aligned}
$$

by the Jacobi identity. The last term is equal to $\varphi([a, b])(x + A)$, as required.

So $\varphi(A)$ is a Lie subalgebra of $\mathsf{gl}(\bar{L})$ and $\dim \varphi(A) < \dim L$. To apply the inductive hypothesis, we need to know that $\varphi(a)$ is a nilpotent linear transformation of $\bar{L}$. But $\varphi(a)$ is induced by $\operatorname{ad} a$; by Lemma 5.1, we know that $\operatorname{ad} a : L \to L$ is nilpotent and therefore $\varphi(a)$ is as well.

By the inductive hypothesis, there is some non-zero element $y + A \in \bar{L}$ such that $\varphi(a)(y + A) = [a, y] + A = 0$ for all $a \in A$. That is, $[a, y] \in A$ for all $a \in A$. Set $\tilde{A} := A \oplus \operatorname{Span}\{y\}$. This is a Lie subalgebra of L containing A. By maximality, $\tilde{A}$ must be equal to L. Therefore $L = A \oplus \operatorname{Span}\{y\}$. As A is an ideal in $\tilde{A}$, it follows that A is an ideal of L.

Step 2: We now apply the inductive hypothesis to $A \subseteq \mathsf{gl}(V)$. This gives us a non-zero $w \in V$ such that $a(w) = 0$ for all $a \in A$. Hence

$$W = \{v \in V : a(v) = 0 \text{ for all } a \in A\}$$

is a non-zero subspace of V.

By Lemma 5.4, W is invariant under L, so in particular $y(W) \subseteq W$. Since y is nilpotent, the restriction of y to W is also nilpotent. Hence there is some non-zero vector $v \in W$ such that $y(v) = 0$. We may write any $x \in L$ in the form $x = a + \beta y$ for some $a \in A$ and some $\beta \in F$. Doing this, we have

$$x(v) = a(v) + \beta y(v) = 0.$$

This shows that v is a non-zero vector in the kernel of every element of L. $\qquad \square$

6.2 Proof of Engel's Theorem

The remainder of the proof is closely analogous to part (ii) of Exercise 6.1. We use induction on $\dim V$. If $V = \{0\}$, then there is nothing to do, so we assume that $\dim V \geq 1$.

By Proposition 6.2, there is a non-zero vector $v \in V$ such that $x(v) = 0$ for all $x \in L$. Let $U = \mathrm{Span}\{v\}$ and let $\bar{V}$ be the quotient vector space V/U. Any $x \in L$ induces a linear transformation $\bar{x}$ of $\bar{V}$. The map $L \to \mathsf{gl}(\bar{V})$ given by $x \mapsto \bar{x}$ is easily checked to be a Lie algebra homomorphism.

The image of L under this homomorphism is a subalgebra of $\mathsf{gl}(\bar{V})$ which satisfies the hypothesis of Engel's Theorem. Moreover, $\dim(\bar{V}) = n - 1$, so by the inductive hypothesis there is a basis of $\bar{V}$ such that with respect to this basis all $\bar{x}$ are strictly upper triangular. If this basis is $\{v_i + U : 1 \leq i \leq n-1\}$, then $\{v, v_1, \ldots, v_{n-1}\}$ is a basis for V. As $x(v) = 0$ for each $x \in L$, the matrices of elements of L with respect to this basis are strictly upper triangular. $\square$

6.3 Another Point of View

We now give another way to think about Engel's Theorem that does not rely on L being given to us as a subalgebra of some $\mathsf{gl}(V)$. Recall that a Lie algebra is said to be nilpotent if for some $m \geq 1$ we have $L^m = 0$ or, equivalently, if for all $x_0, x_1, \ldots, x_m \in V$ we have

$$[x_0, [x_1, \ldots, [x_{m-1}, x_m] \ldots]] = 0.$$

Theorem 6.3 (Engel's Theorem, second version)

A Lie algebra L is nilpotent if and only if for all $x \in L$ the linear map $\mathrm{ad}\,x : L \to L$ is nilpotent.

Proof

We begin by proving the easier "only if" direction. Note that L is nilpotent if and only if there is some $m \geq 1$ such that

$$\mathrm{ad}\,x_0 \circ \mathrm{ad}\,x_1 \ldots \circ \mathrm{ad}\,x_{m-1} = 0$$

for all $x_0, x_1, \ldots, x_{m-1} \in L$, so with this m we have $(\mathrm{ad}\,x)^m = 0$ for all $x \in L$ and so every $\mathrm{ad}\,x$ is nilpotent.

We now prove the "if" direction. Let $\bar{L} = \text{ad}\, L$ denote the image of L under the adjoint homomorphism, $\text{ad} : L \to \mathsf{gl}(L)$. By hypothesis, every element of $\bar{L}$ is a nilpotent linear transformation of L, so by Engel's Theorem, in its original version, there is a basis of L in which every $\text{ad}\, x$ is strictly upper triangular. It now follows from Exercise 5.4 that $\bar{L}$ is nilpotent. Finally, Lemma 4.9(b) implies that L is nilpotent. $\qquad\qquad\qquad\qquad\qquad\qquad\qquad\qquad\qquad\qquad\qquad\Box$

Remark 6.4 (A trap in Engel's Theorem)

It is very tempting to assume that a Lie subalgebra L of $\mathsf{gl}(V)$ is nilpotent if and only if there is a basis of V such that the elements of L are represented by strictly upper triangular matrices.

However, the "only if" direction is false. For example, any 1-dimensional Lie algebra is (trivially) nilpotent. Let I denote the identity map in $\mathsf{gl}(V)$. The Lie subalgebra Span $\{I\}$ of $\mathsf{gl}(V)$ is therefore nilpotent. In any basis of V, the map I is represented by the identity matrix, which is certainly not strictly upper triangular.

Engel's Theorem has many applications, both in the proofs to come, and in more general linear algebra. We give a few such applications in the exercises at the end of this chapter.

6.4 Lie's Theorem

Let L be a Lie subalgebra of $\mathsf{gl}(V)$. We would now like to understand when there is a basis of V such that the elements of L are all represented by upper triangular matrices. The answer is given below in Lie's Theorem.

Theorem 6.5 (Lie's Theorem)

Let V be an n-dimensional complex vector space and let L be a solvable Lie subalgebra of $\mathsf{gl}(V)$. Then there is a basis of V in which every element of L is represented by an upper triangular matrix.

The following exercise outlines a proof of the corresponding result for a single linear transformation.

Exercise 6.2

Let V be an n-dimensional complex vector space where $n \geq 1$ and let $x : V \to V$ be a linear map.

(i) Show that x has an eigenvector $v \in V$.

(ii) Let $U = \mathrm{Span}\,\{v\}$. Show that x induces a linear transformation $\bar{x} : V/U \to V/U$. By induction, we know that there is a basis $\{v_1 + U, \ldots, v_{n-1} + U\}$ of V/U such that $\bar{x}$ has an upper triangular matrix. Prove that $\{v, v_1, \ldots, v_{n-1}\}$ is a basis of V and that the matrix of x in this basis is upper triangular.

As in the proof of Engel's Theorem, the main step is the generalisation of part (i).

Proposition 6.6

Let V be a non-zero complex vector space. Suppose that L is a solvable Lie subalgebra of $\mathsf{gl}(V)$. Then there is some non-zero $v \in V$ which is a simultaneous eigenvector for all $x \in L$.

Proof

This looks similar to Proposition 6.2, but this time we need the full power of the Invariance Lemma, as non-zero eigenvalues are not quite as convenient as zero eigenvalues for calculations.

As before, we use induction on $\dim(L)$. If $\dim(L) = 1$, the result for a single linear transformation gives us all we need, so we may assume that $\dim(L) > 1$. Since L is solvable, we know that L' is properly contained in L. Choose a subspace A of L which contains L' and is such that $L = A \oplus \mathrm{Span}\{z\}$ for some $0 \neq z \in L$.

By Lemma 4.1 and Example 2.4, A is an ideal of L, and by Lemma 4.4(a) (which says that a subalgebra of a solvable algebra is solvable) A is solvable. We may now apply the inductive hypothesis to obtain a vector $w \in V$ which is a simultaneous eigenvector for all $a \in A$. Let $\lambda : A \to \mathbf{C}$ be the corresponding weight, so $a(w) = \lambda(a)w$ for all $a \in A$. Let V_λ be the weight space corresponding to λ:

$$V_\lambda = \{v \in V : a(v) = \lambda(a)v \text{ for all } a \in A\}.$$

This eigenspace is non-zero, as it contains w. By the Invariance Lemma (Lemma 5.5), the space V_λ is L-invariant. Hence there is some non-zero $v \in V_\lambda$ which is an eigenvector for z.

We claim that v is a simultaneous eigenvector for all $x \in L$. Any $x \in L$ may be written in the form $x = a + \beta z$ for some $a \in A$ and $\beta \in \mathbf{C}$. We have

$$x(v) = a(v) + \beta z(v) = \lambda(a)v + \beta\mu v$$

where μ is the eigenvalue of z corresponding to v. This completes the proof. $\square$

The remainder of the proof of Lie's Theorem is completely analogous to the proof of Engel's Theorem given in §6.2, so we leave this to the reader.

Remark 6.7 (Generalisations of Lie's Theorem)

One might ask whether Lie's theorem holds for more general fields. As the eigenvalues of a matrix in upper triangular form are its diagonal entries, we will certainly need the field to contain the eigenvalues of all the elements of $\mathsf{gl}(L)$. The simplest way to achieve this is to require that our field be algebraically closed.

The example in Exercise 6.4 below shows that we also need our field to have characteristic zero. This is perhaps more surprising. We used this assumption in the last step of the proof of the Invariance Lemma when we deduced from $m\,\mathrm{tr}([x, y]) = 0$ that $\mathrm{tr}([x, y]) = 0$. (This would of course be inadmissible if the characteristic of the field divided m.)

EXERCISES

6.3.   Let L be a complex Lie algebra. Show that L is nilpotent if and only if every 2-dimensional subalgebra of L is abelian. (Use the second version of Engel's Theorem.)

6.4.   Let p be prime and let F be a field of characteristic p. Consider the following $p \times p$ matrices:

$$x = \begin{pmatrix} 0 & 1 & 0 & \cdots & 0 \\ 0 & 0 & 1 & \cdots & 0 \\ \vdots & \vdots & \vdots & \ddots & \vdots \\ 0 & 0 & 0 & \cdots & 1 \\ 1 & 0 & 0 & \cdots & 0 \end{pmatrix}, \quad y = \begin{pmatrix} 0 & 0 & \cdots & 0 & 0 \\ 0 & 1 & \cdots & 0 & 0 \\ \vdots & \vdots & \ddots & \vdots & \vdots \\ 0 & 0 & \cdots & p{-}2 & 0 \\ 0 & 0 & \cdots & 0 & p{-}1 \end{pmatrix}.$$

Check that $[x, y] = x$. Deduce that x and y span a 2-dimensional solvable subalgebra L of $\mathsf{gl}(p, F)$. Show that x and y have no common eigenvector, and so the conclusions of Proposition 6.6 and Lie's Theorem fail. Show that the conclusion of part (i) of the following exercise also fails.

6.5.†   (i) Let L be a solvable Lie subalgebra of $\mathsf{gl}(V)$, where V is a complex vector space. Show that every element of L' is a nilpotent endomorphism of V.

     (ii) Let L be a complex Lie algebra. Show that L is solvable if and only if L' is nilpotent.

6.6.   Use Lie's Theorem to give another proof of Proposition 5.7.

7

Some Representation Theory

7.1 Definitions

In this chapter, we introduce the reader to representations of Lie algebras. Our aim is to examine the ways in which an abstract Lie algebra can be viewed, concretely, as a subalgebra of the endomorphism algebra of a finite-dimensional vector space. Representations are defined as follows.

Definition 7.1

Let L be a Lie algebra over a field F. A *representation* of L is a Lie algebra homomorphism

$$\varphi : L \to \mathsf{gl}(V)$$

where V is a finite-dimensional vector space over F. For brevity, we will sometimes omit mention of the homomorphism and just say that V is a representation of L.

As well as being of intrinsic interest, representations of Lie algebras arise frequently in applications of Lie theory to other areas of mathematics and physics. Moreover, we shall see that representations provide a very good way to understand the structure of Lie algebras.

If V is a representation of a Lie algebra L over a field F, then we can fix a basis of V and write the linear transformations of V afforded by elements of L as matrices. Alternatively, we can specify a representation by giving a homomorphism $L \to \mathsf{gl}(n, F)$. In this setup, a representation is sometimes called a *matrix representation*.

Suppose that $\varphi : L \to \mathsf{gl}(V)$ is a representation. By Exercise 1.6, the image of φ is a Lie subalgebra of $\mathsf{gl}(V)$ and the kernel of φ is an ideal of L. By working with $\varphi(L)$, we will in general lose some information about L. If, however, the kernel is zero, or equivalently the map φ is one-to-one, then no information is lost. In this case, the representation is said to be *faithful*.

7.2 Examples of Representations

(1) We have already encountered the adjoint map

$$\mathrm{ad} : L \to \mathsf{gl}(L), \quad (\mathrm{ad}\, x)y = [x, y].$$

This is a Lie homomorphism (see §1.4) so ad provides a representation of L, with $V = L$. This representation is known as the *adjoint representation*. It occurs very often and encapsulates much of the structure of L. We saw in §1.4 that the kernel of the adjoint representation is $Z(L)$. Hence the adjoint representation is faithful precisely when the centre of L is zero. For example, this happens when $L = \mathsf{sl}(2, \mathbf{C})$.

Exercise 7.1

Consider the adjoint representation of $\mathsf{sl}(2, \mathbf{C})$. Show that with respect to the basis (h, e, f) given in Exercise 1.12, ad h has matrix

$$\begin{pmatrix} 0 & 0 & 0 \\ 0 & 2 & 0 \\ 0 & 0 & -2 \end{pmatrix}.$$

Find the matrices representing ad e and ad f.

(2) Suppose that L is a Lie subalgebra of $\mathsf{gl}(V)$. The inclusion map $L \to \mathsf{gl}(V)$ is trivially a Lie algebra homomorphism; for example, because it is the restriction to of the identity map on $\mathsf{gl}(V)$ to L. The corresponding representation is known as the *natural representation* of L.

We have seen several such representations, mostly in the form of matrices, for example, $\mathsf{sl}(n, \mathbf{C})$, or the upper triangular matrices $\mathsf{b}(n, \mathbf{C})$. The Lie algebras considered in Exercise 3.3 all have natural representations. The natural representation is always faithful.

(3) Every Lie algebra has a *trivial* representation. To define this representation, take $V = F$ and and define $\varphi(x) = 0$ for all $x \in L$. This representation is never faithful for non-zero L.

(4) The Lie algebra $\mathbf{R}^3_\wedge$ introduced in §1.2 has a trivial centre, and so its adjoint representation is a 3-dimensional faithful representation. In Exercise 1.15 you were asked to show that $\mathbf{R}^3_\wedge$ is isomorphic to a subalgebra of $\mathsf{gl}(\mathbf{R}^3)$; this gives another 3-dimensional representation. In fact, these representations are in a sense the same; we make this sense more precise in §7.6 below.

7.3 Modules for Lie Algebras

Definition 7.2

Suppose that L is a Lie algebra over a field F. A *Lie module* for L, or alternatively an *L-module*, is a finite-dimensional F-vector space V together with a map

$$L \times V \to V \quad (x, v) \mapsto x \cdot v$$

satisfying the conditions

$$(\lambda x + \mu y) \cdot v = \lambda(x \cdot v) + \mu(y \cdot v), \tag{M1}$$

$$x \cdot (\lambda v + \mu w) = \lambda(x \cdot v) + \mu(x \cdot w), \tag{M2}$$

$$[x, y] \cdot v = x \cdot (y \cdot v) - y \cdot (x \cdot v), \tag{M3}$$

for all $x, y \in L$, $v, w \in V$, and $\lambda, \mu \in F$.

For example, if V is a vector space and L is a Lie subalgebra of $\mathsf{gl}(V)$, then one can readily verify that V is an L-module, where $x \cdot v$ is the image of v under the linear map x.

Note that (M1) and (M2) are equivalent to saying that the map $(x, v) \mapsto x \cdot v$ is bilinear. Condition (M2) implies that for each $x \in L$ the map $v \mapsto x \cdot v$ is a linear endomorphism of V, so elements of L act on V by linear maps. The significance of (M3) will be revealed shortly.

Lie modules and representations are two different ways to describe the same structures. Given a representation $\varphi : L \to \mathsf{gl}(V)$, we may make V an L-module by defining

$$x \cdot v := \varphi(x)(v) \quad \text{for } x \in L, \, v \in V.$$

To show that this works, we must check that the axioms for an L-module are satisfied.

(M1) We have

$$(\lambda x + \mu y) \cdot v = \varphi(\lambda x + \mu y)(v) = (\lambda\varphi(x) + \mu\varphi(y))(v)$$

as φ is linear. By the definition of addition and scalar multiplication of linear maps, this is $\lambda\varphi(x)(v) + \mu\varphi(y)(v) = \lambda(x \cdot v) + \mu(y \cdot v)$.

(M2) Condition M2 is similarly verified:

$$x \cdot (\lambda v + \mu w) = \varphi(x)(\lambda v + \mu w) = \lambda\varphi(x)(v) + \mu\varphi(x)(w) = \lambda(x \cdot v) + \mu(x \cdot w).$$

(M3) By our definition and because φ is a Lie homomorphism, we have

$$[x, y] \cdot v = \varphi([x, y])(v) = [\varphi(x), \varphi(y)](v).$$

As the Lie bracket in $\mathsf{gl}(V)$ is the commutator of linear maps, this equals

$$\varphi(x)\big(\varphi(y)(v)\big) - \varphi(y)\big(\varphi(x)(v)\big) = x \cdot (y \cdot v) - y \cdot (x \cdot v).$$

Conversely, if V is an L-module, then we can regard V as a representation of L. Namely, define

$$\varphi : L \to \mathsf{gl}(V)$$

by letting $\varphi(x)$ be the linear map $v \mapsto x \cdot v$.

Exercise 7.2

Check that φ is a Lie algebra homomorphism.

Remark 7.3

It would be reasonable to ask at this point why we have introduced both representations and L-modules. The reason is that both approaches have their advantages, and sometimes one approach seems more natural than the other. For modules, the notation is easier, and some of the concepts can appear more natural. On the other hand, having an explicit homomorphism to work with can be helpful when we are more interested in the Lie algebra than in the vector space on which it acts.

A similar situation arises when we have a group acting on a set. Here one can choose between the equivalent languages of G-actions and permutation representations. Again, both have their advantages.

7.4 Submodules and Factor Modules

Suppose that V is a Lie module for the Lie algebra L. A *submodule* of V is a subspace W of V which is invariant under the action of L. That is, for each $x \in L$ and for each $w \in W$, we have $x \cdot w \in W$. In the language of representations, submodules are known as *subrepresentations*.

Example 7.4

Let L be a Lie algebra. We may make L into an L-module via the adjoint representation. The submodules of L are exactly the ideals of L. (You are asked to check this in Exercise 7.5 below.)

Example 7.5

Let $L = \mathsf{b}(n, F)$ be the Lie algebra of $n \times n$ upper triangular matrices and let V be the natural L-module, so by definition $V = F^n$ and the action of L is given by applying matrices to column vectors.

Let $e_1, \ldots, e_n$ be the standard basis of F^n. For $1 \le r \le n$, let $W_r = \mathrm{Span}\{e_1, \ldots, e_r\}$. Exercise 7.6 below asks you to show that W_r is a submodule of V.

Example 7.6

Let L be a complex solvable Lie algebra. Suppose that $\varphi : L \to \mathsf{gl}(V)$ is a representation of L. As φ is a homomorphism, $\mathrm{im}\, \varphi$ is a solvable subalgebra of $\mathsf{gl}(V)$. Proposition 6.6 (the main step in the proof of Lie's Theorem) implies that V has a 1-dimensional subrepresentation.

Suppose that W is a submodule of the L-module V. We can give the quotient vector space V/W the structure of an L-module by setting

$$x \cdot (v + W) := (x \cdot v) + W \quad \text{for } x \in L \text{ and } v \in V.$$

We call this module the *quotient* or *factor module V/W*.

As usual, we must first check that the action of L is well-defined. Suppose that $v + W = v' + W$. Then $(x \cdot v) + W - (x \cdot v') + W = x \cdot (v - v') + W = 0 + W$ as $v - v' \in W$ and W is L-invariant. We should also check that the action satisfies the three conditions (M1), (M2) and (M3). We leave this to the reader. She will see that each property follows easily from the corresponding property of the L-module V.

Example 7.7

Suppose that I is an ideal of the Lie algebra L. We have seen that I is a submodule of L when L is considered as an L-module via the adjoint representation. The factor module L/I becomes an L-module via

$$x \cdot (y + I) := (\mathrm{ad}\, x)(y) + I = [x, y] + I.$$

We can interpret this in a different way. We know that L/I is also a Lie algebra (see §2.2), with the Lie bracket given by

$$[x + I, y + I] = [x, y] + I.$$

So, regarded as an L/I-module, the factor module L/I is the adjoint representation of L/I on itself.

Example 7.8

Let $L = \mathsf{b}(n, F)$ and $V = F^n$ as in Example 7.5 above. Fix r between 1 and n and let $W = V_r$ be the r-dimensional submodule defined in the example.

Let $x \in L$ have matrix X with respect to the standard basis. The matrix for the action of x on W with respect to the basis $e_1, \ldots, e_r$ is obtained by taking the upper left $r \times r$ block of X. Moreover, the matrix for the action of x on the quotient space V/W with respect to the basis $e_{r+1} + W, \ldots, e_n + W$ is obtained by taking the lower right $n - r \times n - r$ block of X:

$$X = \begin{pmatrix} a_{11} & a_{12} & \cdots & a_{1r} & & & \\ 0 & a_{22} & \cdots & a_{2r} & & \star & \\ \vdots & \vdots & \ddots & \vdots & & & \\ 0 & 0 & \cdots & a_{rr} & & & \\ & & & & a_{r\,r+1} & \cdots & a_{rn} \\ & & 0 & & \vdots & \ddots & \vdots \\ & & & & 0 & \cdots & a_{nn} \end{pmatrix}.$$

As usual, $\star$ marks a block of unimportant entries.

7.5 Irreducible and Indecomposable Modules

The Lie module V is said to be *irreducible*, or *simple*, if it is non-zero and it has no submodules other than $\{0\}$ and V.

Suppose that V is a non-zero L-module. We may find an irreducible submodule S of V by taking any non-zero submodule of V of minimal dimension. (If V is irreducible, then we just take V itself.) The quotient module V/S will itself have an irreducible submodule, S' and so on. In a sense V is made up of the simple modules $S, S', \ldots$. One says that irreducible modules are the *building blocks* for all finite-dimensional modules.

Example 7.9

(1) If V is 1-dimensional, then V is irreducible. For example, the trivial representation is always irreducible.

(2) If L is a simple Lie algebra, then L viewed as an L-module via the adjoint representation is irreducible. For example $\mathsf{sl}(2, \mathbf{C})$ is irreducible as an $\mathsf{sl}(2, \mathbf{C})$-module.

(3) If L is a complex solvable Lie algebra then it follows from Example 7.6 that all the irreducible representations of L are 1-dimensional.

Given a module V, how can one determine whether or not it is irreducible? One useful criterion is given in the following exercise.

Exercise 7.3

Show that V is irreducible if and only if for any non-zero $v \in V$ the submodule generated by v contains all elements of V. The submodule generated by v is defined to be the subspace of V spanned by all elements of the form

$$x_1 \cdot (x_2 \cdot \ldots \cdot (x_m \cdot v) \ldots),$$

where $x_1, \ldots, x_m \in L$.

Another criterion that is sometimes useful is given in Exercise 7.13 at the end of this chapter.

If V is an L-module such that $V = U \oplus W$, where both U and W are L-submodules of V, we say that V is the *direct sum* of the L-module U and W. The module V is said to be *indecomposable* if there are no non-zero submodules U and W such that $V = U \oplus W$. Clearly an irreducible module is indecomposable. The converse does not usually hold: See the second example below.

The L-module V is *completely reducible* if it can be written as a direct sum of irreducible L-modules; that is, $V = S_1 \oplus S_2 \oplus \ldots \oplus S_k$, where each S_i is an irreducible L-module.

Example 7.10

(1) Let F be a field and let $L = \mathsf{d}(n, F)$ be the subalgebra of $\mathsf{gl}(n, F)$ consisting of diagonal matrices. The natural module $V = F^n$ is completely reducible. If $S_i = \mathrm{Span}\,\{e_i\}$, then S_i is a 1-dimensional simple submodule of V and $V = S_1 \oplus \ldots \oplus S_n$. As the S_i are the weight spaces for L, we can view this as a reformulation of Example 5.2 in Chapter 5.

(2) If $L = \mathsf{b}(n, F)$ where F is a field, then the natural module $V = F^n$ is indecomposable: See Exercise 7.6 below. Note, however, that provided $n \geq 2$, V is not irreducible since $\mathrm{Span}\,\{e_1\}$ is a non-trivial submodule. So V is not completely reducible.

7.6 Homomorphisms

Let L be a Lie algebra and let V and W be L-modules. An *L-module homomorphism* or *Lie homomorphism* from V to W is a linear map $\theta : V \to W$ such that

$$\theta(x \cdot v) = x \cdot \theta(v) \quad \text{for all } v \in V \text{ and } x \in L.$$

An isomorphism is a bijective L-module homomorphism.

Let $\varphi_V : L \to \mathsf{gl}(V)$ and $\varphi_W : L \to \mathsf{gl}(W)$ be the representations corresponding to V and W. In the language of representations, the condition becomes

$$\theta \circ \varphi_V = \varphi_W \circ \theta.$$

Homomorphisms are in particular linear maps, so we can talk about the kernel and image of an L-module homomorphism. And as expected there are the following isomorphism theorems for L-modules.

Theorem 7.11 (Isomorphism Theorems)

(a) Let $\theta : V \to W$ be a homomorphism of L-modules. Then $\ker \theta$ is an L-submodule of V and $\mathrm{im}\,\theta$ is an L-submodule of W, and there is an isomorphism of L-modules,

$$V/\ker \theta \cong \mathrm{im}\,\theta.$$

(b) If U and W are submodules of V, then $U + W$ and $U \cap W$ are submodules of V and $(U + W)/W \cong U/U \cap W$.

(c) If U and W are submodules of V such that $U \subseteq W$, then W/U is a submodule of V/U and the factor module $(V/U)/(W/U)$ is isomorphic to V/W.

Exercise 7.4

Prove the isomorphism theorems for modules by adapting the argument used to prove the corresponding theorems for Lie algebras (see Theorem 2.2).

Finally, if U is a submodule of V, then there is a bijective correspondence between the submodules of V/U and the submodules of V containing U. Again this is precisely analogous to the corresponding result for Lie algebras, given in §2.3.

Example 7.12

Let L be the 1-dimensional abelian Lie algebra spanned, say, by x. We may define a representation of L on a vector space V by mapping x to *any* element of $\mathsf{gl}(V)$. Let W be another vector space. The representations of L corresponding to linear maps $f : V \to V$ and $g : W \to W$ are isomorphic if and only if there is a vector space isomorphism $\theta : V \to W$ such that $\theta f = g\theta$, or equivalently, such that $\theta f \theta^{-1} = g$. Thus the representations are isomorphic if and only if there are bases of V and W in which f and g are represented by the same matrix.

For example, the 2-dimensional matrix representations defined by

$$x \mapsto \begin{pmatrix} 3 & -2 \\ 1 & 0 \end{pmatrix} \quad \text{and} \quad x \mapsto \begin{pmatrix} 1 & 0 \\ 0 & 2 \end{pmatrix}$$

are isomorphic because the two matrices are conjugate (as may be checked by diagonalising the first). For a more substantial example, see Exercise 7.9.

7.7 Schur's Lemma

One of the best ways to understand the structure of a module for a Lie algebra is to look at the homomorphisms between it and other modules. It is natural to begin by looking at the homomorphisms between irreducible modules.

Suppose that S and T are irreducible Lie modules and that $\theta : S \to T$ is a non-zero module homomorphism. Then $\operatorname{im} \theta$ is a non-zero submodule of T, so $\operatorname{im} \theta = T$. Similarly, $\ker \theta$ is a proper submodule of S, so $\ker \theta = 0$. It follows that θ is an isomorphism from S to T, so there are *no* non-zero homomorphisms between non-isomorphic irreducible modules.

Now we consider the homomorphism from an irreducible module to itself.

Lemma 7.13 (Schur's Lemma)

Let L be a complex Lie algebra and let S be a finite-dimensional irreducible L-module. A map $\theta : S \to S$ is an L-module homomorphism if and only if θ is a scalar multiple of the identity transformation; that is, $\theta = \lambda 1_S$ for some $\lambda \in \mathbf{C}$

Proof

The "if" direction should be clear. For the "only if" direction, suppose that $\theta : S \to S$ is an L-module homomorphism. Then θ is, in particular, a linear map of a complex vector space, and so it must have an eigenvalue, say λ. Now $\theta - \lambda 1_S$ is also an L-module homomorphism. The kernel of this map contains a λ-eigenvector for θ, and so it is a non-zero submodule of S. As S is irreducible, $S = \ker(\theta - \lambda 1_S)$; that is, $\theta = \lambda 1_S$.   □

Schur's Lemma has many applications, for example the following.

Lemma 7.14

Let L be a complex Lie algebra and let V be an irreducible L-module. If $z \in Z(L)$, then z acts by scalar multiplication on V; that is, there is some $\lambda \in \mathbf{C}$ such that $z \cdot v = \lambda v$ for all $v \in V$.

Proof

The map $v \mapsto z \cdot v$ is an L-module homomorphism, for if $x \in L$ then

$$z \cdot (x \cdot v) = x \cdot (z \cdot v) + [z, x] \cdot v = x \cdot (z \cdot v)$$

since $[z, x] = 0$. Now apply Schur's Lemma.   □

A corollary of the previous lemma is that the simple modules for an Abelian Lie algebra over $\mathbf{C}$ are 1-dimensional. We deduce this as follows. Suppose that V is a simple module for the abelian Lie algebra L. By this lemma, every element of L acts by scalar multiplication on V, so any non-zero $v \in V$ spans a 1-dimensional submodule of V. As V is irreducible, this submodule must be all of V.

Remark 7.15

In Chapters 5 and 6, we worked with the hypothesis that a Lie algebra L was a subalgebra of $\mathbf{gl}(V)$ for some vector space V. In other words, we were assuming that L had a faithful representation on V. Although we shall not prove it in this book, it is a theorem (Ado's Theorem) that every Lie algebra has a faithful representation.

If L is a Lie algebra and $\varphi : L \to \mathsf{gl}(V)$ is a representation (not necessarily faithful), then we can apply the results of Chapters 5 and 6 to the subalgebra $\operatorname{im} \varphi \subseteq \mathsf{gl}(V)$. In particular, it makes sense to talk about weights and weight spaces for L.

EXERCISES

7.5. Let L be a finite-dimensional Lie algebra. Let V be L with the L-module structure on V given by the adjoint representation of L. Show that the submodules of V are precisely the ideals of L.

7.6. Let F be a field and let $L = \mathsf{b}(n, F)$ and $V = F^n$.

(i) Check that V is an L-module, where the structure is given by the natural representation; that is, by applying matrices to column vectors.

(ii) Let $e_1, \ldots, e_n$ be the standard basis of F^n. For $1 \le r \le n$, let $W_r = \operatorname{Span}\{e_1, \ldots, e_r\}$. Prove that W_r is a submodule of V.

(iii) Show that every non-zero submodule of V is equal to one of the W_r. Deduce that each W_r is indecomposable and that if $n \ge 2$ then V is not completely reducible as an L-module.

7.7. This exercise generalises the remarks made in Example 7.8. Let L be a Lie algebra, and let V be a finite-dimensional L-module with a submodule W of dimension m. By taking a basis of V which contains a basis of W, show that V has a basis in which the action of every $x \in L$ is represented by a "block matrix"

$$\begin{pmatrix} X_1 & X_2 \\ 0 & X_3 \end{pmatrix},$$

where X_1 is a square matrix of size $m \times m$. Show that X_1 is the matrix of x restricted to W and that X_3 represents the action of x on the factor module V/W.

7.8.† Let L be the Heisenberg algebra with basis f, g, z such that $[f, g] = z$ and z is central. Show that L does not have a faithful finite-dimensional irreducible representation.

7.9.†   Let L be the 2-dimensional complex non-abelian Lie algebra found in §3.1. We showed that L has a basis x, y such that $[x, y] = x$. Check that we may define a representation of L on $\mathbf{C}^2$ by setting

$$\varphi(x) = \begin{pmatrix} 0 & 1 \\ 0 & 0 \end{pmatrix}, \quad \varphi(y) = \begin{pmatrix} -1 & 1 \\ 0 & 0 \end{pmatrix}.$$

Show that this representation is isomorphic to the adjoint representation of L on itself.

7.10.*   We now attempt to classify all 2-dimensional complex representations of the two-dimensional non-abelian Lie algebra L. (The notation is as in the previous exercise.)

(i) Suppose that V is a 2-dimensional representation of L which is not faithful. Show that then x acts as zero on V and that V is completely described by the action of y. Deduce that there are as many such representations as there are similarity classes of 2×2 complex matrices.

(ii) Suppose that V is a faithful two-dimensional representation of L. We know from Example 7.9(3) that V has a 1-dimensional irreducible submodule spanned, say, by v. Extend v to a basis of V, say by w.

  (a) Show that the matrix of x with respect to this basis is of the form

$$\begin{pmatrix} 0 & b \\ 0 & 0 \end{pmatrix},$$

where b is non-zero. By replacing v with bv we may assume from now on that $b = 1$.

  (b) Show that the matrix of y with respect to the basis bv, w is of the form

$$\begin{pmatrix} \lambda & c \\ 0 & \mu \end{pmatrix},$$

where $\mu - \lambda = 1$.

  (c) Conversely, check that letting y act as such a matrix really does define a 2-dimensional faithful representation of L.

  (d) Show that if λ is non-zero then there is a basis of V extending v in which y is represented by a diagonal matrix. Hence show that two representations in which y does not kill the submodule $x(V)$ are isomorphic if and only if the matrices representing y have the same trace.

(e) Classify up to isomorphism all representations in which y acts as zero on the submodule $x(V)$.

(iii) If V is a 2-dimensional module for L, then so is its dual module $V^\star$ (dual modules are defined in Exercise 7.12 below). Where does it appear in the classification?

7.11. Let L be a Lie algebra over a field F. Suppose that $\varphi : L \to \mathrm{gl}(1, F)$ is a 1-dimensional representation of L. Show that $\varphi(L') = 0$.

Show that any representation of L/L' can be viewed as a representation of L on which L' acts trivially.

When $F = \mathbf{C}$, show that if $L' \neq L$ then L has infinitely many non-isomorphic 1-dimensional modules, but if $L' = L$ then the only 1-dimensional representation of L is the trivial representation.

7.12. Direct sums are not the only way in which we can construct new modules from old.

(i) Let V be a module for the Lie algebra L. Show that we may make the dual space $V^\star$ into an L-module by defining

$$(x \cdot \theta)(v) = -\theta(x \cdot v) \quad \text{for } x \in L,\ \theta \in V^\star,\ v \in V.$$

Show that the adjoint representation of $\mathbf{R}^3_\wedge$ is self-dual. Prove more generally that V is isomorphic to $V^\star$ if and only if there is a basis of V in which the matrices representing the action of L are all skew-symmetric.

(ii) Let V and W be L-modules. Show that $\mathrm{Hom}(V, W)$, the vector space of linear maps from V to W, may be made into an L-module by defining

$$(x \cdot \theta)(v) = x \cdot (\theta(v)) - \theta(x \cdot v)$$

for $x \in L$, $\theta \in \mathrm{Hom}(V, W)$, and $v \in V$. Show that the linear map $\theta \in \mathrm{Hom}(V, W)$ is an L-module homomorphism if and only if $x \cdot \theta = 0$ for all $x \in L$.

7.13.† Let L be a Lie algebra and let A be an abelian subalgebra of L. Suppose that the L-module V decomposes as a direct sum of weight spaces for A. Show that any L-submodule of V has a basis of common eigenvectors for A.

(This exercise gives a partial converse to the Invariance Lemma (Lemma 5.5). Recall that this lemma states that if A is an *ideal* of L, then any weight space for A is L-invariant.)

Representations of $\mathsf{sl}(2, \mathbf{C})$

In this chapter, we study the finite-dimensional irreducible representations of $\mathsf{sl}(2, \mathbf{C})$. In doing this, we shall see, in a stripped-down form, many of the ideas needed to study representations of an arbitrary semisimple Lie algebra. Later we will see that representations of $\mathsf{sl}(2, \mathbf{C})$ control a large part of the structure of all semisimple Lie algebras.

We shall use the basis of $\mathsf{sl}(2, \mathbf{C})$ introduced in Exercise 1.12 throughout this chapter. Recall that we set

$$e = \begin{pmatrix} 0 & 1 \\ 0 & 0 \end{pmatrix}, \; f = \begin{pmatrix} 0 & 0 \\ 1 & 0 \end{pmatrix}, \; h = \begin{pmatrix} 1 & 0 \\ 0 & -1 \end{pmatrix}.$$

8.1 The Modules V_d

We begin by constructing a family of irreducible representations of $\mathsf{sl}(2, \mathbf{C})$.

Consider the vector space $\mathbf{C}[X, Y]$ of polynomials in two variables X, Y with complex coefficients. For each integer $d \geq 0$, let V_d be the subspace of homogeneous polynomials in X and Y of degree d. So V_0 is the 1-dimensional vector space of constant polynomials, and for $d \geq 1$, the space V_d has as a basis the monomials $X^d, X^{d-1}Y, \ldots, XY^{d-1}, Y^d$. This basis shows that V_d has dimension $d + 1$ as a $\mathbf{C}$-vector space.

We now make V_d into an $\mathsf{sl}(2, \mathbf{C})$-module by specifying a Lie algebra homomorphism $\varphi : \mathsf{sl}(2, \mathbf{C}) \to \mathsf{gl}(V_d)$. Since $\mathsf{sl}(2, \mathbf{C})$ is linearly spanned by the matrices e, f, h, the map φ will be determined once we have specified $\varphi(e)$, $\varphi(f)$, $\varphi(h)$.

We let

$$\varphi(e) := X\frac{\partial}{\partial Y};$$

that is, $\varphi(e)$ is the linear map which first differentiates a polynomial with respect to Y and then multiplies it with X. This preserves the degrees of polynomials and so maps V_d into V_d. Similarly, we let

$$\varphi(f) := Y\frac{\partial}{\partial X}.$$

Finally, we let

$$\varphi(h) := X\frac{\partial}{\partial X} - Y\frac{\partial}{\partial Y}.$$

Notice that

$$\varphi(h)(X^a Y^b) = (a - b)X^a Y^b,$$

so h acts diagonally on V_d with respect to our chosen basis.

Theorem 8.1

With these definitions, φ is a representation of sl(2, **C**).

Proof

By construction, φ is linear. Thus, all we have to check is that φ preserves Lie brackets. By linearity, it is enough to check this on the basis elements of sl(2, **C**), so there are just three equations we need to verify.

(1) We begin by showing $[\varphi(e), \varphi(f)] = \varphi([e, f]) = \varphi(h)$. If we apply the left-hand side to a basis vector $X^a Y^b$ with $a, b \geq 1$ and $a + b = d$, we get

$$\begin{aligned}
[\varphi(e), \varphi(f)](X^a Y^b) &= \varphi(e)\left(\varphi(f)(X^a Y^b)\right) - \varphi(f)\left(\varphi(e)(X^a Y^b)\right) \\
&= \varphi(e)\left(aX^{a-1}Y^{b+1}\right) - \varphi(f)\left(bX^{a+1}Y^{b-1}\right) \\
&= a(b+1)X^a Y^b - b(a+1)X^a Y^b \\
&= (a - b)X^a Y^b.
\end{aligned}$$

This is the same as $\varphi(h)(X^a Y^b)$. We check separately the action on X^d,

$$\begin{aligned}
[\varphi(e), \varphi(f)](X^d) &= \varphi(e)\left(\varphi(f)(X^d)\right) - \varphi(f)\left(\varphi(e)(X^d)\right) \\
&= \varphi(e)\left(dX^{d-1}Y\right) - \varphi(f)(0) = dX^d,
\end{aligned}$$

which is the same as $\varphi(h)(X^d)$. Similarly, one checks the action on Y^d, so $[\varphi(e), \varphi(f)]$ and $\varphi(h)$ agree on a basis of V_d and so are the same linear map.

(2) We also need $[\varphi(h), \varphi(e)] = \varphi([h, e]) = \varphi(2e) = 2\varphi(e)$. Again we can prove this by applying the maps to basis vectors of V_d. For $b \geq 1$, we get

$$
\begin{aligned}
[\varphi(h), \varphi(e)](X^a Y^b) &= \varphi(h)\left(\varphi(e)(X^a Y^b)\right) - \varphi(e)\left(\varphi(h)(X^a Y^b)\right) \\
&= \varphi(h)\left(b X^{a+1} Y^{b-1}\right) - \varphi(e)\left((a-b) X^a Y^b\right) \\
&= b\left((a+1) - (b-1)\right) X^{a+1} Y^{b-1} - (a-b) b X^{a+1} Y^{b-1} \\
&= 2b X^{a+1} Y^{b-1}.
\end{aligned}
$$

This is the same as $2\varphi(e)(X^a Y^b)$. If $b = 0$ and $a = d$, then a separate verification is needed. We leave this to the reader.

(3) Similarly, one can check that $[\varphi(h), \varphi(f)] = -2\varphi(f)$. Again, we leave this to the reader.

$\square$

8.1.1 Matrix Interpretation

It can be useful to know the matrices that correspond to the action of e, f, h on V_d; these give the matrix representation corresponding to φ.

As usual, we take the basis $X^d, X^{d-1}Y, \ldots, Y^d$ of V_d. The calculations in the proof of Theorem 8.1 show that the matrix of $\varphi(e)$ with respect to this basis is

$$
\begin{pmatrix}
0 & 1 & 0 & \cdots & 0 \\
0 & 0 & 2 & \cdots & 0 \\
\vdots & \vdots & \vdots & \ddots & \vdots \\
0 & 0 & 0 & \cdots & d \\
0 & 0 & 0 & \cdots & 0
\end{pmatrix},
$$

the matrix of $\varphi(f)$ is

$$
\begin{pmatrix}
0 & 0 & \cdots & 0 & 0 \\
d & 0 & \cdots & 0 & 0 \\
0 & d-1 & \cdots & 0 & 0 \\
\vdots & \vdots & \ddots & \vdots & \vdots \\
0 & 0 & \cdots & 1 & 0
\end{pmatrix},
$$

and $\varphi(h)$ is diagonal:

$$
\begin{pmatrix}
d & 0 & \cdots & 0 & 0 \\
0 & d-2 & \cdots & 0 & 0 \\
\vdots & \vdots & \ddots & \vdots & \vdots \\
0 & 0 & \cdots & -d+2 & 0 \\
0 & 0 & \cdots & 0 & -d
\end{pmatrix}
$$

where the diagonal entries are the numbers $d - 2k$, where $k = 0, 1, \ldots, d$. By explicitly computing the commutators of these matrices, we can give another (but equivalent) way to prove that φ is a representation of sl(2, **C**).

Another way to represent the action of h, e, f is to draw a diagram like

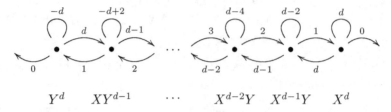

where loops represent the action of h, arrows to the right represent the action of e, and arrows to the left represent the action of f.

8.1.2 Irreducibility

One virtue of the diagram above is that it makes it almost obvious that the sl(2, **C**)-submodule of V_d generated by any particular basis element $X^a Y^b$ contains all the basis elements and so is all of V_d.

Exercise 8.1

Check this assertion.

A possible disadvantage of our diagram is that it may blind us to the existence of the *many other* vectors in V_d, which, while linear combinations of the basis vectors, are not basis vectors themselves.

Theorem 8.2

The sl(2, **C**)-module V_d is irreducible.

Proof

Suppose U is a non-zero sl(2, **C**)-submodule of V_d. Then $h \cdot u \in U$ for all $u \in U$. Since h acts diagonalisably on V_d, it also acts diagonalisably on U, so there is an eigenvector of h which lies in U. We have seen that all eigenspaces of h on V_d are one-dimensional, and each eigenspace is spanned by some monomial $X^a Y^b$, so the submodule U must contain some monomial, and by the exercise above, U contains a basis for V_d. Hence $U = V_d$. $\qquad \square$

8.2 Classifying the Irreducible sl(2, C)-Modules

It is clear that for different d the sl(2, **C**)-modules V_d cannot be isomorphic, as they have different dimensions. In this section, we prove that any finite-dimensional irreducible sl(2, **C**)-module is isomorphic to one of the V_d. Our strategy will be to look at the eigenvectors and eigenvalues of h. For brevity, we shall write $e^2 \cdot v$ rather than $e \cdot (e \cdot v)$, and so on.

Lemma 8.3

Suppose that V is an sl(2, **C**)-module and $v \in V$ is an eigenvector of h with eigenvalue λ.

(i) Either $e \cdot v = 0$ or $e \cdot v$ is an eigenvector of h with eigenvalue $\lambda + 2$.

(ii) Either $f \cdot v = 0$ or $f \cdot v$ is an eigenvector of h with eigenvalue $\lambda - 2$.

Proof

As V is a representation of sl(2, **C**), we have

$$h \cdot (e \cdot v) = e \cdot (h \cdot v) + [h, e] \cdot v = e \cdot (\lambda v) + 2e \cdot v = (\lambda + 2)e \cdot v.$$

The calculation for $f \cdot v$ is similar.   □

Lemma 8.4

Let V be a finite-dimensional sl(2, **C**)-module. Then V contains an eigenvector w for h such that $e \cdot w = 0$.

Proof

As we work over **C**, the linear map $h : V \to V$ has at least one eigenvalue and so at least one eigenvector. Let $h \cdot v = \lambda v$. Consider the vectors

$$v, \ e \cdot v, \ e^2 \cdot v, \ldots.$$

If they are non-zero, then by Lemma 8.3 they form an infinite sequence of h-eigenvectors with distinct eigenvalues. Eigenvectors with different eigenvalues are linearly independent, so V would contain infinitely many linearly independent vectors, a contradiction.

Therefore there exists $k \geq 0$ such that $e^k \cdot v \neq 0$ and $e^{k+1} \cdot v = 0$. If we set $w = e^k \cdot v$, then $h \cdot w = (\lambda + 2k)w$ and $e \cdot w = 0$.   □

We are now ready to prove our main result.

Theorem 8.5

If V is a finite-dimensional irreducible sl(2, **C**)-module, then V is isomorphic to one of the V_d.

Proof

By Lemma 8.4, V has an h-eigenvector w such that $e \cdot w = 0$. Suppose that $h \cdot w = \lambda w$. Consider the sequence of vectors

$$w, \ f \cdot w, \ f^2 \cdot w, \ldots.$$

By the proof of Lemma 8.4, there exists $d \geq 0$ such that $f^d \cdot w \neq 0$ and $f^{d+1} \cdot w = 0$.

Step 1: We claim that the vectors $w, f \cdot w, \ldots, f^d \cdot w$ form a basis for a submodule of V. They are linearly independent because, by Lemma 8.3, they are eigenvectors for h with distinct eigenvalues. By construction, the span of $w, f \cdot w, \ldots, f^d \cdot w$ is invariant under h and f. To show that it is invariant under e, we shall prove by induction on k that

$$e \cdot (f^k \cdot w) \in \mathrm{Span}\{f^j \cdot w : 0 \leq j < k\}.$$

If $k = 0$, then we know that $e \cdot w = 0$. For the inductive step, note that

$$e \cdot (f^k \cdot w) = (fe + h) \cdot (f^{k-1} \cdot w).$$

By the inductive hypothesis, $e \cdot (f^{k-1} \cdot w)$ is in the span of the $f^j \cdot w$ for $j < k-1$ and therefore $f e f^{k-1} \cdot w$ is in the span of all $f^j \cdot w$ for $j < k$. Moreover $h f^{k-1} \cdot w$ is a scalar multiple of $f^{k-1} \cdot w$. This gives the inductive step.

Now, since V is irreducible, the submodule spanned by the $f^k \cdot w$ for $0 \leq k \leq d$ is equal to V.

Step 2: In this step, we shall show that $\lambda = d$. The matrix of h with respect to the basis $w, f \cdot w, \ldots, f^d \cdot w$ of V is diagonal, with trace

$$\lambda + (\lambda - 2) + \ldots + (\lambda - 2d) = (d+1)\lambda - (d+1)d.$$

Since $[e, f] = h$, the matrix of h is equal to the commutator of the matrices of e and f, so it has trace zero and $\lambda = d$.

Step 3: To finish, we produce an explicit isomorphism $V \cong V_d$. As we have seen, V has basis $\{w, f \cdot w, \ldots, f^d \cdot w\}$. Furthermore, V_d has basis

$$\{X^d, f \cdot X^d, \ldots, f^d \cdot X^d\},$$

where $f^k \cdot X^d$ is a scalar multiple of $X^{d-k} Y^k$. Moreover, the eigenvalue of h on $f^k \cdot w$ is the same as the eigenvalue of h on $f^k \cdot X^d$. Clearly, to have a homomorphism, we must have a map which takes h-eigenvectors to h-eigenvectors for the same eigenvalue. So we may set

$$\psi(w) = X^d$$

and then we must define ψ by

$$\psi(f^k \cdot w) := f^k \cdot X^d.$$

This defines a vector space isomorphism, which commutes with the actions of f and h. To show that it also commutes with the action of e, we use induction on k and a method similar to Step 1. Explicitly, for $k = 0$ we have $\psi(e \cdot w) = 0$ and $e\psi(w) = e \cdot X^d = 0$. For the inductive step,

$$\psi(ef^k \cdot w) = \psi((fe + h) \cdot (f^{k-1} \cdot w)) = f \cdot \psi(ef^{k-1} \cdot w) + h \cdot \psi(f^{k-1} \cdot w)$$

using that ψ commutes with f and h. We use the inductive hypothesis to take e out and obtain that the expression can be written as

$$(fe + h) \cdot \psi(f^{k-1} \cdot w) = ef \cdot \psi(f^{k-1} \cdot w) = e \cdot \psi(f^k \cdot w). \qquad \square$$

Corollary 8.6

If V is a finite-dimensional representation of sl(2, **C**) and $w \in V$ is an h-eigenvector such that $e \cdot w = 0$, then $h \cdot w = dw$ for some non-negative integer d and the submodule of V generated by w is isomorphic to V_d.

Proof

Step 1 in the previous proof shows that for some $d \geq 0$ the vectors $w, f \cdot w, \ldots, f^d \cdot w$ span a submodule of V. Now apply steps 2 and 3 to this submodule to get the required conclusions. $\qquad \square$

A vector v of the type considered in this corollary is known as a *highest-weight vector*. If d is the associated eigenvalue of h, then d is said to be a *highest weight*. (See §15.1 for a more general setting.)

8.3 Weyl's Theorem

In Exercise 7.6, we gave an example of a module for a Lie algebra that was not completely reducible; that is, it could not be written as a direct sum of irreducible submodules. Finite-dimensional representations of complex semisimple Lie algebras however are much better behaved.

Theorem 8.7 (Weyl's Theorem)

Let L be a complex semisimple Lie algebra. Every finite-dimensional representation of L is completely reducible.

The proof of Weyl's Theorem is fairly long, so we defer it to Appendix B. Weyl's Theorem tells us that to understand the finite-dimensional representations of a semisimple Lie algebra it is sufficient to understand its irreducible representations. We give an introduction to this topic in §15.1.

In the main part of this book, we shall only need to apply Weyl's Theorem to representations of sl(2, **C**), in which case a somewhat easier proof, exploiting properties of highest-weight vectors, is possible. Exercise 8.6 in this chapter does the groundwork, and the proof is finished in Exercise 9.15. (Both of these exercises have solutions in Appendix E.)

EXERCISES

8.2.   Find explicit isomorphisms between

   (i) the trivial representation of sl(2, **C**) and V_0;

   (ii) the natural representation of sl(2, **C**) and V_1;

   (iii) the adjoint representation of sl(2, **C**) and V_2.

8.3.   Show that the subalgebra of sl(3, **C**) consisting of matrices of the form

$$\begin{pmatrix} \star & \star & 0 \\ \star & \star & 0 \\ 0 & 0 & 0 \end{pmatrix}$$

is isomorphic to sl(2, **C**). We may therefore regard sl(3, **C**) as a module for sl(2, **C**), with the action given by $x \cdot y = [x, y]$ for $x \in$ sl(2, **C**) and $y \in$ sl(3, **C**). Show that as an sl(2, **C**)-module

$$\text{sl}(3, \mathbf{C}) \cong V_2 \oplus V_1 \oplus V_1 \oplus V_0.$$

8.4.   Suppose that V is a finite-dimensional module for $\mathsf{sl}(2, \mathbf{C})$. Show, by using Weyl's Theorem and the classification of irreducible representations in this chapter, that V is determined up to isomorphism by the eigenvalues of h. In particular, prove that if V is the direct sum of k irreducible modules, then

$$k = \dim W_0 + \dim W_1,$$

where $W_r = \{v \in V : h \cdot v = rv\}$.

8.5.   Let V be an $\mathsf{sl}(2, \mathbf{C})$-module, not necessarily finite-dimensional. Suppose $w \in V$ is a highest-weight vector of weight λ; that is, $e \cdot w = 0$ and $h \cdot w = \lambda w$ for some $\lambda \in \mathbf{C}$, and $w \neq 0$. Show that

(i) for $k = 1, 2, \ldots$ we have $e \cdot (f^k \cdot w) = k(\lambda - k + 1)f^{k-1} \cdot w$, and

(ii) $e^k f^k \cdot w = (k!)^2 \binom{\lambda}{k}w$.

Deduce that if $\binom{\lambda}{k} \neq 0$ then the set of all $f^j \cdot w$ for $0 \leq j \leq k$ is linearly independent. Hence show that if V is finite-dimensional, then λ must be a non-negative integer.

8.6.† Let M be a finite-dimensional $\mathsf{sl}(2, \mathbf{C})$-module. Define a linear map $c : M \longrightarrow M$ by

$$c(v) = \left(ef + fe + \frac{1}{2}h^2\right) \cdot v \quad \text{for } v \in M.$$

(i) Show that c is a homomorphism of $\mathsf{sl}(2, \mathbf{C})$-modules. *Hint:* For example, to show that c commutes with the action of e, show that $(efe + fe^2 + \frac{1}{2}h^2 e) \cdot v$ and $(e^2 f + efe + \frac{1}{2}eh^2) \cdot v$ are both equal to $(2efe + \frac{1}{2}heh) \cdot v$.

(ii) By Schur's Lemma, c must act as a scalar, say λ_d, on the irreducible module V_d. Show that $\lambda_d = \frac{1}{2}d(d + 2)$, and deduce that d is determined by λ_d.

(iii) Let $\lambda_1, \ldots, \lambda_r$ be the distinct eigenvalues of c acting on M. Let the primary decomposition of M be

$$M = \bigoplus_{i=1}^{r} \ker(c - \lambda_i 1_M)^{m_i}.$$

Show that the summands are $\mathsf{sl}(2, \mathbf{C})$-submodules.

So, to express the module as a direct sum of simple modules, we therefore may assume that M has just one generalised eigenspace, where c has, say, eigenvalue λ.

(iv) Let U be an irreducible submodule of M. Suppose that U is isomorphic to V_d. Show by considering the action of c on V_d that $\lambda = \frac{1}{2}d(d+2)$ and hence that *any* irreducible submodule of M is isomorphic to V_d.

(v) Show more generally that if N is a submodule of M, then any irreducible submodule of M/N is isomorphic to V_d.

The linear map c is known as the *Casimir operator*. The following exercise gives an indication of how it was first discovered; it will appear again in the proof of Weyl's Theorem (see Appendix B).

8.7.   Exercise 1.14 gives a way to embed the real Lie algebra $\mathbf{R}^3_\wedge$ into sl(2, **C**). With the given solution, we would take

$$\psi(x) = \begin{pmatrix} 0 & 1/2 \\ -1/2 & 0 \end{pmatrix}, \psi(y) = \begin{pmatrix} 0 & -i/2 \\ -i/2 & 0 \end{pmatrix}, \psi(z) = \begin{pmatrix} -i/2 & 0 \\ 0 & i/2 \end{pmatrix}$$

Check that $\psi(x)^2 + \psi(y)^2 + \psi(z)^2 = -3/4I$, where I is the 2×2 identity matrix. By expressing x, y, z in terms of e, f, h, recover the description of the Casimir operator given above.

The interested reader might like to look up "angular momentum" or "Pauli matrices" in a book on quantum mechanics to see the physical interpretation of the Casimir operator.

Cartan's Criteria

How can one decide whether a complex Lie algebra is semisimple? Working straight from the definition, one would have to test every single ideal for solvability, seemingly a daunting task. In this chapter, we describe a practical way to decide whether a Lie algebra is semisimple or, at the other extreme, solvable, by looking at the traces of linear maps.

We have already seen examples of the usefulness of taking traces. For example, we made an essential use of the trace map when proving the Invariance Lemma (Lemma 5.5). An important identity satisfied by trace is

$$\operatorname{tr}([a, b]c) = \operatorname{tr}(a[b, c])$$

for linear transformations a, b, c of a vector space. This holds because $\operatorname{tr} b(ac) = \operatorname{tr}(ac)b$; we shall see its usefulness in the course of this chapter. Furthermore, note that a nilpotent linear transformation has trace zero.

From now on, we work entirely over the complex numbers.

9.1 Jordan Decomposition

Working over the complex numbers allows us to consider the Jordan normal form of linear transformations. We use this to define for each linear transformation x of a complex vector space V a unique *Jordan decomposition*. The Jordan decomposition of x is the unique expression of x as a sum $x = d + n$ where $d : V \to V$ is diagonalisable, $n : V \to V$ is nilpotent, and d and n commute.

Very often, a diagonalisable linear map of a complex vector space is also said to be *semisimple*.

We review the Jordan normal form and prove the existence and uniqueness of the Jordan decomposition in Appendix A. The lemma below is also proved in this Appendix; see §16.6.

Lemma 9.1

Let x be a linear transformation of the complex vector space V. Suppose that x has Jordan decomposition $x = d + n$, where d is diagonalisable, n is nilpotent, and d and n commute.

(a) There is a polynomial $p(X) \in \mathbf{C}[X]$ such that $p(x) = d$.

(b) Fix a basis of V in which d is diagonal. Let $\bar{d}$ be the linear map whose matrix with respect to this basis is the complex conjugate of the matrix of d. There is a polynomial $q(X) \in \mathbf{C}[X]$ such that $q(x) = \bar{d}$.   □

Using Jordan decomposition, we can give a concise reinterpretation of two earlier results (see Exercise 1.17 and Lemma 5.1).

Exercise 9.1

Let V be a vector space, and suppose that $x \in \mathsf{gl}(V)$ has Jordan decomposition $d+n$. Show that $\operatorname{ad} x : \mathsf{gl}(V) \to \mathsf{gl}(V)$ has Jordan decomposition $\operatorname{ad} d + \operatorname{ad} n$.

9.2 Testing for Solvability

Let V be a complex vector space and let L be a Lie subalgebra of $\mathsf{gl}(V)$. Why might it be reasonable to expect solvability to be visible from the traces of the elements of L? The following exercise (which repeats part of Exercise 6.5) gives one indication.

Exercise 9.2

Suppose that L is solvable. Use Lie's Theorem to show that there is a basis of V in which every element of L' is represented by a strictly upper triangular matrix. Conclude that $\operatorname{tr} xy = 0$ for all $x \in L$ and $y \in L'$.

Thus we have a necessary condition, in terms of traces, for L to be solvable. Remarkably, this condition is also sufficient. Before proving this, we give a small example.

Example 9.2

Let L be the 2-dimensional non-abelian Lie algebra with basis x, y such that $[x, y] = x$, which we constructed in §3.1. In this basis we have

$$\operatorname{ad} x = \begin{pmatrix} 0 & 1 \\ 0 & 0 \end{pmatrix}, \quad \operatorname{ad} y = \begin{pmatrix} -1 & 0 \\ 0 & 0 \end{pmatrix}.$$

As expected, $\operatorname{tr} \operatorname{ad} x = 0$.

Proposition 9.3

Let V be a complex vector space and let L be a Lie subalgebra of $\mathsf{gl}(V)$. If $\operatorname{tr} xy = 0$ for all $x, y \in L$, then L is solvable.

Proof

We shall show that every $x \in L'$ is a nilpotent linear map. It will then follow from Engel's Theorem (Theorem 6.1) that L' is nilpotent, and so, by the 'if' part of Exercise 6.5(ii), L is solvable.

Let $x \in L'$ have Jordan decomposition $x = d + n$, where d is diagonalisable, n is nilpotent, and d and n commute. We may fix a basis of V in which d is diagonal and n is strictly upper triangular. Suppose that d has diagonal entries $\lambda_1, \ldots, \lambda_m$. Since our aim is to show that $d = 0$, it will suffice to show that

$$\sum_{i=1}^{m} \lambda_i \bar{\lambda}_i = 0.$$

The matrix of $\bar{d}$ is diagonal, with diagonal entries $\bar{\lambda}_i$ for $1 \le i \le m$. A simple computation shows that

$$\operatorname{tr} \bar{d} x = \sum_{i=1}^{m} \lambda_i \bar{\lambda}_i.$$

Now, as $x \in L'$, we may express x as a linear combination of commutators $[y, z]$ with $y, z \in L$, so we need to show that $\operatorname{tr}(\bar{d}[y, z]) = 0$. By the identity mentioned at the start of this chapter, this is equivalent to

$$\operatorname{tr}([\bar{d}, y]z) = 0.$$

This will hold by our hypothesis, provided we can show that $[\bar{d}, y] \in L$. In other words, we must show that $\operatorname{ad} \bar{d}$ maps L into L.

By Exercise 9.1, the Jordan decomposition of $\operatorname{ad} x$ is $\operatorname{ad} d + \operatorname{ad} n$. Therefore, by part (b) of Lemma 9.1, there is a polynomial $p(X) \in \mathbf{C}[X]$ such that $p(\operatorname{ad} x) = \overline{\operatorname{ad} d} = \operatorname{ad} \bar{d}$. Now $\operatorname{ad} x$ maps L into itself, so $p(\operatorname{ad} x)$ does also. $\square$

To apply this proposition to an abstract Lie algebra L, we need a way to regard L as a subalgebra of some $\mathsf{gl}(V)$. The adjoint representation of L is well-suited to this purpose, as L is solvable if and only if $\operatorname{ad} L$ is solvable.

Theorem 9.4

Let L be a complex Lie algebra. Then L is solvable if and only if $\operatorname{tr}(\operatorname{ad} x \circ \operatorname{ad} y) = 0$ for all $x \in L$ and all $y \in L'$.

Proof

Suppose that L is solvable. Then $\operatorname{ad} L \subseteq \mathsf{gl}(L)$ is a solvable subalgebra of $\mathsf{gl}(L)$, so the result now follows from Exercise 9.2.

Conversely, if $\operatorname{tr}(\operatorname{ad} x \circ \operatorname{ad} y) = 0$ for all $x \in L$ and all $y \in L'$, then Proposition 9.3 implies that $\operatorname{ad} L'$ is solvable. So L' is solvable, and hence L is solvable. □

9.3 The Killing Form

Definition 9.5

Let L be a complex Lie algebra. The *Killing form* on L is the symmetric bilinear form defined by

$$\kappa(x, y) := \operatorname{tr}(\operatorname{ad} x \circ \operatorname{ad} y) \quad \text{for } x, y \in L.$$

The Killing form is bilinear because ad is linear, the composition of maps is bilinear, and tr is linear. (The reader may wish to write out a more careful proof of this.) It is symmetric because $\operatorname{tr} ab = \operatorname{tr} ba$ for linear maps a and b. Another very important property of the Killing form is its *associativity*, which states that for all $x, y, z \in L$ we have

$$\kappa([x, y], z) = \kappa(x, [y, z]).$$

This follows from the identity for trace mentioned at the start of this chapter.

Using the Killing form, we can state Theorem 9.4 as follows.

Theorem 9.6 (Cartan's First Criterion)

The complex Lie algebra L is solvable if and only if $\kappa(x, y) = 0$ for all $x \in L$ and $y \in L'$. □

Example 9.7

Let L be the 2-dimensional non-abelian Lie algebra with basis x, y such that $[x, y] = x$. The matrices computed in Example 9.2 show that $\kappa(x, x) = \kappa(x, y) = \kappa(y, x) = 0$ and $\kappa(y, y) = 1$. The matrix of κ in the basis x, y is therefore

$$\begin{pmatrix} 0 & 0 \\ 0 & 1 \end{pmatrix}.$$

The Killing form is compatible with restriction to ideals. Suppose that L is a Lie algebra and I is an ideal of L. We write κ for the Killing form on L and κ_I for the Killing form on I, considered as a Lie algebra in its own right.

Lemma 9.8

If $x, y \in I$, then $\kappa_I(x, y) = \kappa(x, y)$.

Proof

Take a basis for I and extend it to a basis of L. If $x \in I$, then $\operatorname{ad} x$ maps L into I, so the matrix of $\operatorname{ad} x$ in this basis is of the form

$$\begin{pmatrix} A_x & B_x \\ 0 & 0 \end{pmatrix},$$

where A_x is the matrix of $\operatorname{ad} x$ restricted to I.

If $y \in I$, then $\operatorname{ad} x \circ \operatorname{ad} y$ has matrix

$$\begin{pmatrix} A_x A_y & A_x B_y \\ 0 & 0 \end{pmatrix},$$

where $A_x \circ A_y$ is the matrix of $\operatorname{ad} x \circ \operatorname{ad} y$ restricted to I. Only the block $A_x A_y$ contributes to the trace of this matrix, so

$$\kappa(x, y) = \operatorname{tr}(A_x B_x) = \kappa_I(x, y). \qquad \square$$

9.4 Testing for Semisimplicity

Recall that a Lie algebra is said to be semisimple if its radical is zero; that is, if it has no non-zero solvable ideals. Since we can detect solvability by using the Killing form, it is perhaps not too surprising that we can also use the Killing form to decide whether or not a Lie algebra is semisimple.

We begin by recalling a small part of the general theory of bilinear forms; for more details, see Appendix A. Let β be a symmetric bilinear form on a finite-dimensional complex vector space V. If S is a subset of V, we define the *perpendicular space* to S by

$$S^\perp = \{x \in V : \beta(x, s) = 0 \text{ for all } s \in S\}.$$

This is a vector subspace of V. We say that β is *non-degenerate* if $V^\perp = 0$; that is, there is no non-zero vector $v \in V$ such that $\beta(v, x) = 0$ for all $x \in V$.

If β is non-degenerate and W is a vector subspace of V, then

$$\dim W + \dim W^\perp = \dim V.$$

Note that even if β is non-degenerate it is possible that $W \cap W^\perp \neq 0$. For example, if κ is the Killing form of $\mathsf{sl}(2, \mathbf{C})$, then $\kappa(e, e) = 0$. (You are asked to compute the Killing form of $\mathsf{sl}(2, \mathbf{C})$ in Exercise 9.4 below.)

Now we specialise to the case where L is a Lie algebra and κ is its Killing form, so perpendicular spaces are taken with respect to κ. We begin with a simple observation which requires the associativity of κ.

Exercise 9.3

Suppose that I is an ideal of L. Show that $I^\perp$ is an ideal of L.

By this exercise, $L^\perp$ is an ideal of L. If $x \in L^\perp$ and $y \in \left(L^\perp\right)'$, then, as in particular $y \in L$, we have $\kappa(x, y) = 0$. Hence it follows from Cartan's First Criterion that $L^\perp$ is a solvable ideal of L. Therefore, if L is semisimple, then $L^\perp = 0$ and κ is non-degenerate.

Again the converse also holds.

Theorem 9.9 (Cartan's Second Criterion)

The complex Lie algebra L is semisimple if and only if the Killing form κ of L is non-degenerate.

Proof

We have just proved the "only if" direction. Suppose that L is not semisimple, so $\operatorname{rad} L$ is non-zero. By Exercise 4.6, L has a non-zero abelian ideal, say A. Take a non-zero element $a \in A$, and let $x \in L$. The composite map

$$\operatorname{ad} a \circ \operatorname{ad} x \circ \operatorname{ad} a$$

sends L to zero, as the image of $\operatorname{ad} x \circ \operatorname{ad} a$ is contained in the abelian ideal A. Hence $(\operatorname{ad} a \circ \operatorname{ad} x)^2 = 0$. Nilpotent maps have trace 0, so $\kappa(a, x) = 0$. This holds for all $x \in L$, so a is a non-zero element in $L^\perp$. Thus κ is degenerate. $\square$

It is possible that $L^\perp$ is properly contained in rad L. For example, Exercise 9.2 shows that this is the case if L is the 2-dimensional non-abelian Lie algebra.

Cartan's Second Criterion is an extremely powerful characterisation of semisimplicity. In our first application, we shall show that a semisimple Lie algebra is a direct sum of simple Lie algebras; this finally justifies the name *semi*simple which we have been using. The following lemma contains the main idea needed.

Lemma 9.10

If I is a non-trivial proper ideal in a complex semisimple Lie algebra L, then $L = I \oplus I^\perp$. The ideal I is a semisimple Lie algebra in its own right.

Proof

As usual, let κ denote the Killing form on L. The restriction of κ to $I \cap I^\perp$ is identically 0, so by Cartan's First Criterion, $I \cap I^\perp = 0$. It now follows by dimension counting that $L = I \oplus I^\perp$.

We shall show that I is semisimple using Cartan's Second Criterion. Suppose that I has a non-zero solvable ideal. By the "only if" direction of Cartan's Second Criterion, the Killing form on I is degenerate. We have seen that the Killing form on I is given by restricting the Killing form on L, so there exists $a \in I$ such that $\kappa(a, x) = 0$ for all $x \in I$. But as $a \in I$, $\kappa(a, y) = 0$ for all $y \in I^\perp$ as well. Since $L = I \oplus I^\perp$, this shows that κ is degenerate, a contradiction. $\square$

We can now prove the following theorem.

Theorem 9.11

Let L be a complex Lie algebra. Then L is semisimple if and only if there are simple ideals $L_1, \ldots, L_r$ of L such that $L = L_1 \oplus L_2 \oplus \ldots \oplus L_r$.

Proof

We begin with the "only if" direction, working by induction on dim L. Let I be an ideal in L of the smallest possible non-zero dimension. If $I = L$, we are done. Otherwise I is a proper simple ideal of L. (It cannot be abelian as by hypothesis L has no non-zero abelian ideals.) By the preceding lemma, $L = I \oplus I^\perp$, where, as an ideal of L, $I^\perp$ is a semisimple Lie algebra of smaller dimension than L.

So, by induction, $I^\perp$ is a direct sum of simple ideals,

$$I^\perp = L_2 \oplus \ldots \oplus L_r.$$

Each L_i is also an ideal of L, as $[I, L_i] \subseteq I \cap I^\perp = 0$, so putting $L_1 = I$ we get the required decomposition.

Now for the "if" direction. Suppose that $L = L_1 \oplus \ldots \oplus L_r$, where the L_r are simple ideals. Let $I = \operatorname{rad} L$; our aim is to show that $I = 0$. For each ideal L_i, $[I, L_i] \subseteq I \cap L_i$ is a solvable ideal of L_i. But the L_i are simple, so

$$[I, L] \subseteq [I, L_1] \oplus \ldots \oplus [I, L_r] = 0.$$

This shows that I is contained in $Z(L)$. But by Exercise 2.6(ii)

$$Z(L) = Z(L_1) \oplus \ldots \oplus Z(L_r).$$

We know that $Z(L_1) = \ldots = Z(L_r) = 0$ as the L_i are simple ideals, so $Z(L) = 0$ and $I = 0$. □

Using very similar ideas, we can prove the following.

Lemma 9.12

If L is a semisimple Lie algebra and I is an ideal of L, then L/I is semisimple.

Proof

We have seen that $L = I \oplus I^\perp$, so L/I is isomorphic to $I^\perp$, which we have seen is a semisimple Lie algebra in its own right. □

9.5 Derivations of Semisimple Lie Algebras

In our next application of Cartan's Second Criterion, we show that the only derivations of a complex semisimple Lie algebra are those of the form $\operatorname{ad} x$ for $x \in L$. More precisely, we have the following.

Proposition 9.13

If L is a finite-dimensional complex semisimple Lie algebra, then $\operatorname{ad} L = \operatorname{Der} L$.

Proof

We showed in Example 1.2 that for each $x \in L$ the linear map $\operatorname{ad} x$ is a derivation of L, so ad is a Lie algebra homomorphism from L to $\operatorname{Der} L$. Moreover, if δ is a derivation of L and $x, y \in L$, then

$$
\begin{aligned}
[\delta, \operatorname{ad} x] y &= \delta[x, y] - \operatorname{ad} x(\delta y) \\
&= [\delta x, y] + [x, \delta y] - [x, \delta y] \\
&= \operatorname{ad}(\delta x) y.
\end{aligned}
$$

Thus the image of $\operatorname{ad} : L \to \operatorname{Der} L$ is an ideal of $\operatorname{Der} L$. This much is true for any Lie algebra.

Now we bring in our assumption that L is complex and semisimple. First, note that $\operatorname{ad} : L \to \operatorname{Der} L$ is one-to-one, as $\ker \operatorname{ad} = Z(L) = 0$, so the Lie algebra $M := \operatorname{ad} L$ is isomorphic to L and therefore it is semisimple as well.

To show that $M = \operatorname{Der} L$, we exploit the Killing form on the Lie algebra $\operatorname{Der} L$. If M is properly contained in $\operatorname{Der} L$ then $M^{\perp} \neq 0$, so it is sufficient to prove that $M^{\perp} = 0$. As M is an ideal of $\operatorname{Der} L$, the Killing form κ_M of M is the restriction of the Killing form on $\operatorname{Der} L$. By Cartan's Second Criterion, κ_M is non-degenerate, so $M^{\perp} \cap M = 0$ and hence $[M^{\perp}, M] = 0$. Thus, if $\delta \in M^{\perp}$ and $\operatorname{ad} x \in M$, then $[\delta, \operatorname{ad} x] = 0$. But we saw above that

$$
[\delta, \operatorname{ad} x] = \operatorname{ad}(\delta x),
$$

so, for all $x \in L$, we have $\delta(x) = 0$; in other words, $\delta = 0$. $\qquad\square$

In Exercise 9.17, this proposition is used to give an alternative proof that a semisimple Lie algebra is a direct sum of simple Lie algebras. Another important application occurs in the following section.

9.6 Abstract Jordan Decomposition

Given a representation $\varphi : L \to \mathbf{gl}(V)$ of a Lie algebra L, we may consider the Jordan decomposition of the linear maps $\varphi(x)$ for $x \in L$.

For a general Lie algebra there is not much that can be said about this decomposition without knowing more about the representation φ. For example, if L is the 1-dimensional abelian Lie algebra, spanned, say by x, then we may define a representation of L on a vector space V by mapping x to *any* element of $\mathbf{gl}(V)$. So the Jordan decomposition of $\varphi(x)$ is essentially arbitrary.

However, representations of a complex semisimple Lie algebra are much better behaved. To demonstrate this, we use derivations to define a Jordan decomposition for elements of an arbitrary complex semisimple Lie algebra. We need the following proposition.

Proposition 9.14

Let L be a complex Lie algebra. Suppose that δ is a derivation of L with Jordan decomposition $\delta = \sigma + \nu$, where σ is diagonalisable and ν is nilpotent. Then σ and ν are also derivations of L.

Proof

For $\lambda \in \mathbf{C}$, let

$$L_\lambda = \{x \in L : (\delta - \lambda 1_L)^m x = 0 \text{ for some } m \geq 1\}$$

be the generalised eigenspace of δ corresponding to λ. Note that if λ is not an eigenvalue of δ, then $L_\lambda = 0$. By the Primary Decomposition Theorem, L decomposes as a direct sum of generalised eigenspaces, $L = \bigoplus_\lambda L_\lambda$, where the sum runs over the eigenvalues of δ. In Exercise 9.8 below, you are asked to show that

$$[L_\lambda, L_\mu] \subseteq L_{\lambda+\mu}.$$

We shall use this to show that σ and ν are derivations.

As σ acts diagonalisably, the λ-eigenspace of σ is L_λ. Take $x \in L_\lambda$ and $y \in L_\mu$. Then, by the above, $[x, y] \in L_{\lambda+\mu}$, so

$$\sigma([x, y]) = (\lambda + \mu)[x, y],$$

which is the same as

$$[\sigma(x), y] + [x, \sigma(y)] = [\lambda x, y] + [x, \mu y].$$

Thus σ is a derivation, and so $\delta - \sigma = \nu$ is also a derivation.   □

Theorem 9.15

Let L be a complex semisimple Lie algebra. Each $x \in L$ can be written uniquely as $x = d + n$, where $d, n \in L$ are such that $\operatorname{ad} d$ is diagonalisable, $\operatorname{ad} n$ is nilpotent, and $[d, n] = 0$. Furthermore, if $y \in L$ commutes with x, then $[d, y] = 0$ and $[n, y] = 0$.

Proof

Let $\operatorname{ad} x = \sigma + \nu$ where $\sigma \in \mathsf{gl}(L)$ is diagonalisable, $\nu \in \mathsf{gl}(L)$ is nilpotent, and $[\sigma, \nu] = 0$. By Proposition 9.14, we know that σ and ν are derivations of the semisimple Lie algebra L. In Proposition 9.13, we saw that $\operatorname{ad} L = \operatorname{Der} L$, so there exist $d, n \in L$ such that $\operatorname{ad} d = \sigma$ and $\operatorname{ad} n = \nu$. As ad is injective and

$$\operatorname{ad} x = \sigma + \nu = \operatorname{ad} d + \operatorname{ad} n = \operatorname{ad}(d + n),$$

we get that $x = d + n$. Moreover, $\operatorname{ad}[d, n] = [\operatorname{ad} d, \operatorname{ad} n] = 0$ so $[d, n] = 0$. The uniqueness of d and n follows from the uniqueness of the Jordan decomposition of $\operatorname{ad} x$.

Suppose that $y \in L$ and that $(\operatorname{ad} x)y = 0$. By Lemma 9.1, σ and ν may be expressed as polynomials in $\operatorname{ad} x$. Let

$$\nu = c_0 1_L + c_1 \operatorname{ad} x + \ldots + c_r (\operatorname{ad} x)^r.$$

Applying ν to y, we see that $\nu(y) = c_0 y$. But ν is nilpotent and $\nu(x) = c_0 x$, so $c_0 = 0$. Thus $\nu(y) = 0$ and so $\sigma(y) = (\operatorname{ad} x - \nu)y = 0$ also. $\square$

We say that x has *abstract Jordan decomposition* $x = d + n$. If $n = 0$, then we say that x is *semisimple*.

There is a potential ambiguity in the terms "Jordan decomposition" and "semisimple" which arises when $L \subseteq \mathsf{gl}(V)$ is a semisimple Lie algebra. In this case, as well as the abstract Jordan decomposition just defined, we may also consider the usual Jordan decomposition, given by taking an element of L and regarding it as a linear map on V. It is an important property of the abstract Jordan decomposition that the two decompositions agree; in particular, an element of L is diagonalisable if and only if it is semisimple.

Take $x \in L$. Suppose that the usual Jordan decomposition of x, as an element of $\mathsf{gl}(V)$, is $d + n$. By Exercise 9.1, the Jordan decomposition of the map $\operatorname{ad} x : L \to L$ is $\operatorname{ad} d + \operatorname{ad} n$, so by definition $d + n$ is also the abstract Jordan decomposition of x.

We are now ready to prove the main result about the abstract Jordan decomposition.

Theorem 9.16

Let L be a semisimple Lie algebra and let $\theta : L \to \mathsf{gl}(V)$ be a representation of L. Suppose that $x \in L$ has abstract Jordan decomposition $x = d + n$. Then the Jordan decomposition of $\theta(x) \in \mathsf{gl}(V)$ is $\theta(x) = \theta(d) + \theta(n)$.

Proof

By Lemma 9.12, $\operatorname{im} \theta \cong L/\ker \theta$ is a semisimple Lie algebra. It therefore makes sense to talk about the abstract Jordan decomposition of elements of $\operatorname{im} \theta$.

Let $x \in L$ have abstract Jordan decomposition $d + n$. It follows from Exercise 9.16 below that the abstract Jordan decomposition of $\theta(x)$, considered as an element of $\operatorname{im} \theta$, is $\theta(d) + \theta(n)$. By the remarks above, this is also the Jordan decomposition of $\theta(x)$, considered as an element of $\operatorname{gl}(V)$. $\qquad \square$

The last theorem is a very powerful result, which we shall apply several times in the next chapter. For another application, see Exercise 9.15 below.

EXERCISES

9.4.† (i) Compute the Killing form of $\operatorname{sl}(2, \mathbf{C})$. This is a symmetric bilinear form on a 3-dimensional vector space, so you should expect it to be described by a symmetric 3×3 matrix. Check that the Killing form is non-degenerate.

(ii) Is the Killing form of $\operatorname{gl}(2, \mathbf{C})$ non-degenerate?

9.5. Suppose that L is a nilpotent Lie algebra over a field F. Show by using the ideals L^m, or otherwise, that the Killing form of L is identically zero. Does the converse hold? (The following exercise may be helpful.)

9.6.† For each of the 3-dimensional complex Lie algebras studied in Chapter 3, find its Killing form with respect to a convenient basis.

9.7. Let $L = \operatorname{gl}(n, \mathbf{C})$. Show that the Killing form of L is given by

$$\kappa(a, b) = 2n \operatorname{tr}(ab) - 2(\operatorname{tr} a)(\operatorname{tr} b).$$

For instance, start with $(\operatorname{ad} b)e_{rs}$, apply $\operatorname{ad} a$, and then express the result in terms of the basis and find the coefficient of e_{rs}. Hence prove that if $n \geq 2$ then $\operatorname{sl}(n, \mathbf{C})$ is semisimple.

9.8. Let δ be a derivation of a Lie algebra L. Show that if $\lambda, \mu \in \mathbf{C}$ and $x, y \in L$, then

$$(\delta - (\lambda + \mu)1_L)^n [x, y] = \sum_{k=0}^{n} \binom{n}{k} \left[(\delta - \lambda 1_L)^k x, (\delta - \mu 1_L)^{n-k} y \right].$$

Hence show that if the primary decomposition of L with respect to δ is $L = \bigoplus_\lambda L_\lambda$ (as in the proof of Proposition 9.14), then

$$[L_\lambda, L_\mu] \subseteq L_{\lambda + \mu}.$$

9.9.   (i) Show that if L is a semisimple Lie algebra then $L' = L$.

     (ii) Suppose that L is the direct sum of simple ideals $L = L_1 \oplus L_2 \oplus \ldots \oplus L_k$. Show that if I is a simple ideal of L, then I is equal to one of the L_i. *Hint*: Consider the ideal $[I, L]$.

     (iii)* If $L' = L$, must L be semisimple?

9.10.   Suppose that L is a Lie algebra over $\mathbf{C}$ and that β is a symmetric, associative bilinear form of L. Show that β induces a linear map

$$\theta : L \to L^*, \quad \theta(x) = \beta(x, -),$$

where by $\beta(x, -)$ we mean the map $y \mapsto \beta(x, y)$. Viewing both L and L^* as L-modules, show that θ is an L-module homomorphism. (The L-module structure of L^* is given by Exercise 7.12.) Deduce that if β is non-degenerate, then L and L^* are isomorphic as L-modules.

9.11.† Let L be a simple Lie algebra over $\mathbf{C}$ with Killing form κ. Use Exercise 9.10 to show that if β is any other symmetric, associative, non-degenerate bilinear form on L, then there exists $0 \neq \lambda \in \mathbf{C}$ such that $\kappa = \lambda\beta$.

9.12.   Assuming that $\mathsf{sl}(n, \mathbf{C})$ is simple, use Exercise 9.11 to show that

$$\kappa(x, y) = 2n\,\mathrm{tr}(xy), \quad x, y \in \mathsf{sl}(n, \mathbf{C}).$$

To identify the scalar λ, it might be useful to take as a standard basis for the Lie algebra; $\{e_{ij} : i \neq j\} \cup \{e_{ii} - e_{i+1,i+1} : 1 \leq i < n\}$.

9.13.   Give an example to show that the condition $[d, n] = 0$ in the Jordan decomposition is necessary. That is, find a matrix x which can be written as $x = d+n$ with d diagonalisable and n nilpotent but where this is not the Jordan decomposition of x.

9.14.† Let L be a complex semisimple Lie algebra. Suppose L has a faithful representation in which $x \in L$ acts diagonalisably. Show that x is a semisimple element of L (in the sense of the abstract Jordan decomposition) and hence that x acts diagonalisably in *any* representation of L.

9.15.†* Suppose that M is an $\mathsf{sl}(2, \mathbf{C})$-module. Use the abstract Jordan decomposition to show that M decomposes as a direct sum of h-eigenspaces. Hence use Exercise 8.6 to show that M is completely reducible.

9.16.† Suppose that L_1 and L_2 are complex semisimple Lie algebras and that $\theta : L_1 \to L_2$ is a surjective homomorphism. Show that if $x \in L_1$ has abstract Jordan decomposition $x = d + n$, then $\theta(x) \in L_2$ has abstract Jordan decomposition $\theta(x) = \theta(d) + \theta(n)$. *Hint:* Exercise 2.8 is relevant.

9.17. Use Exercise 2.13 and Proposition 9.13 (that if L is a complex semisimple Lie algebra, then $\operatorname{ad} L = \operatorname{Der} L$) to give an alternative proof of Theorem 9.11 (that a complex semisimple Lie algebra is a direct sum of simple ideals).

9.18.* Some small-dimensional examples suggest that if L is a Lie algebra and I is an ideal of L, then one can always find a basis of I and extend it to a basis of L in such a way that the Killing form of L has a matrix of the form

$$\begin{pmatrix} \kappa_I & 0 \\ 0 & \star \end{pmatrix}.$$

Is this always the case?

The Root Space Decomposition

We are now ready to embark on the classification of the complex semisimple Lie algebras. So far we have proved the simplicity of only one family of Lie algebras, namely the algebras $\mathsf{sl}(n, \mathbf{C})$ for $n \geq 2$ (see Exercise 9.7). There is, however, a strong sense in which their behaviour is typical of all complex semisimple Lie algebras. We therefore begin by looking at the structures of $\mathsf{sl}(2, \mathbf{C})$ and $\mathsf{sl}(3, \mathbf{C})$ in the belief that this will motivate the strategy adopted in this chapter.

In §3.2.4, we proved that $\mathsf{sl}(2, \mathbf{C})$ was the unique 3-dimensional semisimple complex Lie algebra by proceeding as follows:

(1) We first showed that if L was a 3-dimensional Lie algebra such that $L = L'$, then there was some $h \in L$ such that $\mathrm{ad}\, h$ was diagonalisable.

(2) We then took a basis of L consisting of eigenvectors for $\mathrm{ad}\, h$ and by finding the structure constants with respect to this basis showed that L was isomorphic to $\mathsf{sl}(2, \mathbf{C})$.

In the case of $\mathsf{sl}(3, \mathbf{C})$, a suitable replacement for the element $h \in \mathsf{sl}(2, \mathbf{C})$ is the 2-dimensional subalgebra H of diagonal matrices in $\mathsf{sl}(3, \mathbf{C})$. One can see directly that $\mathsf{sl}(3, \mathbf{C})$ decomposes into a direct sum of common eigenspaces for the elements of $\mathrm{ad}\, H$. Suppose $h \in H$ has diagonal entries a_1, a_2, a_3. Then

$$[h, e_{ij}] = (a_i - a_j)e_{ij}$$

so the elements e_{ij} for $i \neq j$ are common eigenvectors for the elements of $\mathrm{ad}\, H$. Moreover, as H is abelian, H is contained in the kernel of every element of $\mathrm{ad}\, H$.

It will be helpful to express this decomposition using the language of weights and weight spaces introduced in Chapter 5. Define $\varepsilon_i : H \to \mathbf{C}$ by $\varepsilon_i(h) = a_i$. We have

$$(\operatorname{ad} h)e_{ij} = (\varepsilon_i - \varepsilon_j)(h)e_{ij}.$$

Here $\varepsilon_i - \varepsilon_j$ is a weight and e_{ij} is in its associated weight space. In fact one can check that if L_{ij} is the weight space for $\varepsilon_i - \varepsilon_j$, that is

$$L_{ij} = \{x \in \mathsf{sl}(3, \mathbf{C}) : (\operatorname{ad} h)x = (\varepsilon_i - \varepsilon_j)(h)x \text{ for all } h \in H\},$$

then we have $L_{ij} = \operatorname{Span}\{e_{ij}\}$ for $i \neq j$. Hence there is a direct sum decomposition

$$\mathsf{sl}(3, \mathbf{C}) = H \oplus \bigoplus_{i \neq j} L_{ij}.$$

The existence of this decomposition can be seen in a more abstract way. Let L be a complex semisimple Lie algebra and let H be an abelian subalgebra of L consisting of semisimple elements. By definition, $\operatorname{ad} h$ is diagonalisable for every $h \in H$. Moreover, as commuting linear maps may be simultaneously diagonalised, H acts diagonalisably on L in the adjoint representation. We may therefore decompose L into a direct sum of weight spaces for the adjoint action of H.

Our strategy is therefore:

(1) to find an abelian Lie subalgebra H of L that consists entirely of semisimple elements; and

(2) to decompose L into weight spaces for the action of $\operatorname{ad} H$ and then exploit this decomposition to determine information about the structure constants of L.

In the following section, we identify the desirable properties of the subalgebra H and prove some preliminary results about the decomposition. We then show that subalgebras H with these desirable properties always exist and complete step (2).

10.1 Preliminary Results

Suppose that L is a complex semisimple Lie algebra containing an abelian subalgebra H consisting of semisimple elements. What information does this give us about L?

We have seen that L has a basis of common eigenvectors for the elements of ad H. Given a common eigenvector $x \in L$, the eigenvalues are given by the associated weight, $\alpha : H \to \mathbf{C}$, defined by

$$(\operatorname{ad} h)x = \alpha(h)x \text{ for all } h \in H.$$

Weights are elements of the dual space H^*. For each $\alpha \in H^*$, let

$$L_\alpha := \{x \in L : [h, x] = \alpha(h)x \text{ for all } h \in H\}$$

denote the corresponding weight space. One of these weight spaces is the zero weight space:

$$L_0 = \{z \in L : [h, z] = 0 \text{ for all } h \in H\}.$$

This is the same as the centraliser of H in L, $C_L(H)$. As H is abelian, we have $H \subseteq L_0$.

Let Φ denote the set of non-zero $\alpha \in H^*$ for which L_α is non-zero. We can write the decomposition of L into weight spaces for H as

$$L = L_0 \oplus \bigoplus_{\alpha \in \Phi} L_\alpha. \qquad (\star)$$

Since L is finite-dimensional, this implies that Φ is finite.

Lemma 10.1

Suppose that $\alpha, \beta \in H^*$. Then

(i) $[L_\alpha, L_\beta] \subseteq L_{\alpha+\beta}$.

(ii) If $\alpha + \beta \neq 0$, then $\kappa(L_\alpha, L_\beta) = 0$.

(iii) The restriction of κ to L_0 is non-degenerate.

Proof

(i) Take $x \in L_\alpha$ and $y \in L_\beta$. We must show that $[x, y]$, if non-zero, is an eigenvector for each ad $h \in H$, with eigenvalue $\alpha(h) + \beta(h)$. Using the Jacobi identity we get

$$\begin{aligned}
[h, [x, y]] = [[h, x], y] + [x, [h, y]] &= [\alpha(h)x, y] + [x, \beta(h)y] \\
&= \alpha(h)[x, y] + \beta(h)[x, y] \\
&= (\alpha + \beta)(h)[x, y].
\end{aligned}$$

(ii) Since $\alpha + \beta \neq 0$, there is some $h \in H$ such that $(\alpha + \beta)(h) \neq 0$. Now, for any $x \in L_\alpha$ and $y \in L_\beta$, we have, using the associativity of the Killing form,

$$\alpha(h)\kappa(x, y) = \kappa([h, x], y) = -\kappa([x, h], y) = -\kappa(x, [h, y]) = -\beta(h)\kappa(x, y),$$

and hence
$$(\alpha + \beta)(h)\kappa(x, y) = 0.$$
Since by assumption $(\alpha + \beta)(h) \neq 0$, we must have $\kappa(x, y) = 0$.

(iii) Suppose that $z \in L_0$ and $\kappa(z, x) = 0$ for all x in L_0. By (ii), we know that $L_0 \perp L_\alpha$ for all $\alpha \neq 0$. If $x \in L$, then by $(\star)$ we can write x as

$$x = x_0 + \sum_{\alpha \in \Phi} x_\alpha$$

with $x_\alpha \in L_\alpha$. By linearity, $\kappa(z, x) = 0$ for all $x \in L$. Since κ is non-degenerate, it follows that $z = 0$, as required. $\square$

Exercise 10.1

Show that if $x \in L_\alpha$ where $\alpha \neq 0$, then $\operatorname{ad} x$ is nilpotent.

If H is small, then the decomposition $(\star)$ is likely to be rather coarse, with few non-zero weight spaces other than L_0. Furthermore, if H is properly contained in L_0, then we get little information about how the elements in L_0 that are not in H act on L. This is illustrated by the following exercise.

Exercise 10.2

Let $L = \mathsf{sl}(n, \mathbf{C})$, where $n \geq 2$, and let $H = \operatorname{Span}\{h\}$, where $h = e_{11} - e_{22}$. Find $L_0 = C_L(H)$, and determine the direct sum decomposition $(\star)$ with respect to H.

We conclude that for the decomposition $(\star)$ of L into weight spaces to be as useful as possible, H should be as large as possible, and ideally we would have $H = L_0 = C_L(H)$.

Definition 10.2

A Lie subalgebra H of a Lie algebra L is said to be a *Cartan subalgebra* (or CSA) if H is abelian and every element $h \in H$ is semisimple, and moreover H is maximal with these properties.

Note that we do not assume L is semisimple in this definition. For example, the subalgebra H of $\mathsf{sl}(3, \mathbf{C})$ considered in the introduction to this chapter is a Cartan subalgebra of $\mathsf{sl}(3, \mathbf{C})$. One straightforward way to see this is to show that $C_{\mathsf{sl}(3,\mathbf{C})}(H) = H$; thus H is not contained in any larger abelian subalgebra of $\mathsf{sl}(3, \mathbf{C})$.

We remark that some texts use a "maximal toral subalgebra" in place of what we have called a Cartan subalgebra. The connection is discussed at the end of Appendix C, where we establish that the two types of algebras are the same.

10.2 Cartan Subalgebras

Let L be a complex semisimple Lie algebra. We shall show that L has a non-zero Cartan subalgebra. We first note that L must contain semisimple elements. If $x \in L$ has Jordan decomposition $x = s + n$, then by Theorem 9.15 both s and n belong to L. If the semisimple part s were always zero, then by Engel's Theorem (in its second version), L would be nilpotent and therefore solvable. Hence we can find a non-zero semisimple element $s \in L$. We can now obtain a non-zero Cartan subalgebra of L by taking any subalgebra which contains s and which is maximal subject to being abelian and consisting of semisimple elements. (Such a subalgebra must exist because L is finite-dimensional.)

We shall now show that if H is a Cartan subalgebra then $H = C_L(H)$. The proof of this statement is slightly technical, so the reader may prefer to defer or skip some of the details. In this case, she should continue reading at §10.3.

Lemma 10.3

Let H be a Cartan subalgebra of L. Suppose that $h \in H$ is such that the dimension of $C_L(h)$ is minimal. Then every $s \in H$ is central in $C_L(h)$, and so $C_L(h) \subseteq C_L(s)$. Hence $C_L(h) = C_L(H)$.

Proof

We shall show that if s is not central in $C_L(h)$, then there is a linear combination of s and h whose centraliser has smaller dimension than $C_L(h)$.

First we construct a suitable basis for L. We start by taking a basis of $C_L(h) \cap C_L(s)$, $\{c_1, \ldots, c_n\}$, say. As s is semisimple and $s \in C_L(h)$, ad s acts diagonalisably on $C_L(h)$. We may therefore extend this basis to a basis of $C_L(h)$ consisting of ad s eigenvectors, say by adjoining $\{x_1, \ldots, x_p\}$. Similarly we may extend $\{c_1, \ldots, c_n\}$ to a basis of $C_L(s)$ consisting of ad h eigenvectors, say by adjoining $\{y_1, \ldots, y_q\}$. We leave it to the reader to check that

$$\{c_1, \ldots, c_n, x_1, \ldots, x_p, y_1, \ldots, y_q\}$$

is a basis of $C_L(h) + C_L(s)$. Finally, as ad h and ad s commute and act diagonalisably on L, we may extend this basis to a basis of L by adjoining simultaneous eigenvectors for ad h and ad s, say $\{w_1, \ldots, w_r\}$.

Note that if $[s, x_j] = 0$ then $x_j \in C_L(s) \cap C_L(h)$, a contradiction. Similarly, one can check that $[h, y_k] \neq 0$. Let $[h, w_l] = \theta_l w_l$ and $[s, w_l] = \sigma_l w_l$. Again we have $\theta_l, \sigma_l \neq 0$ for $1 \leq l \leq r$. The following table summarises the eigenvalues of $\operatorname{ad} s$, $\operatorname{ad} h$, and $\operatorname{ad} s + \lambda \operatorname{ad} h$, where $\lambda \neq 0$:

	c_i	x_j	y_k	w_l
$\operatorname{ad} s$	0	$\neq 0$	0	σ_l
$\operatorname{ad} h$	0	0	$\neq 0$	θ_l
$\operatorname{ad} s + \lambda \operatorname{ad} h$	0	$\neq 0$	$\neq 0$	$\sigma_l + \lambda \theta_l$

Thus, if we choose λ so that $\lambda \neq 0$ and $\lambda \neq -\sigma_l/\theta_l$ for any l, then we will have

$$C_L(s + \lambda h) = C_L(s) \cap C_L(h).$$

By hypothesis, $C_L(h) \not\subseteq C_L(s)$, so this subspace is of smaller dimension than $C_L(h)$; this contradicts the choice of h.

Now, since $C_L(H)$ is the intersection of the $C_L(s)$ for $s \in H$, it follows that $C_L(h) \subseteq C_L(H)$. The other inclusion is obvious, so we have proved that $C_L(h) = C_L(H)$. $\qquad\square$

Theorem 10.4

If H is a Cartan subalgebra of L and $h \in H$ is such that $C_L(h) = C_L(H)$, then $C_L(h) = H$. Hence H is self-centralising.

Proof

Since H is abelian, H is certainly contained in $C_L(h)$. Suppose $x \in C_L(h)$ has abstract Jordan decomposition $x = s + n$. As x commutes with h, Theorem 9.15 implies that both s and n lie in $C_L(h)$, so we must show that $s \in H$ and $n = 0$.

We almost know already that $s \in H$. Namely, since $C_L(h) = C_L(H)$, we have that s commutes with every element of H and therefore $H + \operatorname{Span}\{s\}$ is an abelian subalgebra of L consisting of semisimple elements. It contains the Cartan subalgebra H and hence by maximality $s \in H$.

To show that the only nilpotent element in $C_L(H)$ is 0 takes slightly more work.

Step 1: $C_L(h)$ is nilpotent. Take $x \in C_L(h)$ with $x = s + n$ as above. Since $s \in H$, it must be central in $C_L(h)$, so, regarded as linear maps from $C_L(h)$ to itself, we have $\operatorname{ad} x = \operatorname{ad} n$. Thus for every $x \in C_L(h)$, $\operatorname{ad} x : C_L(h) \to C_L(h)$ is nilpotent. It now follows from the second version of Engel's Theorem (Theorem 6.3) that $C_L(h)$ is a nilpotent Lie algebra.

Step 2: Every element in $C_L(h)$ is semisimple. Let $x \in C_L(h)$ have abstract Jordan decomposition $x = s + n$ as above. As $C_L(h)$ is nilpotent, it is certainly solvable, so by Lie's Theorem (Theorem 6.5) there is a basis of L in which the maps $\operatorname{ad} x$ for $x \in C_L(h)$ are represented by upper triangular matrices. As $\operatorname{ad} n : L \to L$ is nilpotent, its matrix must be strictly upper triangular. Therefore

$$\kappa(n, z) = \operatorname{tr}(\operatorname{ad} n \circ \operatorname{ad} z) = 0$$

for all $z \in C_L(h)$. By Lemma 10.1(iii), the restriction of κ to $C_L(H)$ is non-degenerate, so we deduce $n = 0$, as required. □

10.3 Definition of the Root Space Decomposition

Let H be a Cartan subalgebra of our semisimple Lie algebra L. As $H = C_L(H)$, the direct sum decomposition of L into weight spaces for H considered in §10.1 may be written as

$$L = H \oplus \bigoplus_{\alpha \in \Phi} L_\alpha,$$

where Φ is the set of $\alpha \in H^\star$ such that $\alpha \neq 0$ and $L_\alpha \neq 0$. Since L is finite-dimensional, Φ is finite.

If $\alpha \in \Phi$, then we say that α is a *root* of L and L_α is the associated *root space*. The direct sum decomposition above is the *root space decomposition*. It should be noted that the roots and root spaces depend on the choice of Cartan subalgebra H.

10.4 Subalgebras Isomorphic to $\mathsf{sl}(2, \mathbf{C})$

We shall now associate to each root $\alpha \in \Phi$ a Lie subalgebra of L isomorphic to $\mathsf{sl}(2, \mathbf{C})$. These subalgebras will enable us to apply the results in Chapter 8 on representations of $\mathsf{sl}(2, \mathbf{C})$ to deduce several strong results on the structure of L. Chapters 11 and 12 give many examples of the theory we develop in the next three sections. See also Exercise 10.6 for a more immediate example.

Lemma 10.5

Suppose that $\alpha \in \Phi$ and that x is a non-zero element in L_α. Then $-\alpha$ is a root and there exists $y \in L_{-\alpha}$ such that $\operatorname{Span}\{x, y, [x, y]\}$ is a Lie subalgebra of L isomorphic to $\mathsf{sl}(2, \mathbf{C})$.

Proof

First we claim that there is some $y \in L_{-\alpha}$ such that $\kappa(x, y) \neq 0$ and $[x, y] \neq 0$. Since κ is non-degenerate, there is some $w \in L$ such that $\kappa(x, w) \neq 0$. Write $w = y_0 + \sum_{\beta \in \Phi} y_\beta$ with $y_0 \in L_0$ and $y_\beta \in L_\beta$. When we expand $\kappa(x, y)$, we find by Lemma 10.1(ii) that the only way a non-zero term can occur is if $-\alpha$ is a root and $y_{-\alpha} \neq 0$, so we may take $y = y_{-\alpha}$. Now, since α is non-zero, there is some $t \in H$ such that $\alpha(t) \neq 0$. For this t, we have

$$\kappa(t, [x, y]) = \kappa([t, x], y) = \alpha(t)\kappa(x, y) \neq 0$$

and so $[x, y] \neq 0$.

Let $S := \mathrm{Span}\{x, y, [x, y]\}$. By Lemma 10.1(i), $[x, y]$ lies in $L_0 = H$. As x and y are simultaneous eigenvectors for all elements of $\mathrm{ad}\, H$, and so in particular for $\mathrm{ad}[x, y]$, this shows that S is a Lie subalgebra of L. It remains to show that S is isomorphic to $\mathsf{sl}(2, \mathbf{C})$.

Let $h := [x, y] \in S$. We claim that $\alpha(h) \neq 0$. If not, then $[h, x] = \alpha(h)x = 0$; similarly $[h, y] = -\alpha(h)y = 0$, so $\mathrm{ad}\, h : L \to L$ commutes with $\mathrm{ad}\, x : L \to L$ and $\mathrm{ad}\, y : L \to L$. By Proposition 5.7, $\mathrm{ad}\, h : L \to L$ is a nilpotent map. On the other hand, because H is a Cartan subalgebra, h is semisimple. The only element of L that is both semisimple and nilpotent is 0, so $h = 0$, a contradiction.

Thus S is a 3-dimensional complex Lie algebra with $S' = S$. By §3.2.4, S is isomorphic to $\mathsf{sl}(2, \mathbf{C})$. $\square$

Using this lemma, we may associate to each $\alpha \in \Phi$ a subalgebra $\mathsf{sl}(\alpha)$ of L isomorphic to $\mathsf{sl}(2, \mathbf{C})$. The following exercise gives a standard basis for this Lie algebra.

Exercise 10.3

Show that for each $\alpha \in \Phi$, $\mathsf{sl}(\alpha)$ has a basis $\{e_\alpha, f_\alpha, h_\alpha\}$ such that

(i) $e_\alpha \in L_\alpha$, $f_\alpha \in L_{-\alpha}$, $h_\alpha \in H$, and $\alpha(h_\alpha) = 2$.

(ii) The map $\theta : \mathsf{sl}(\alpha) \to \mathsf{sl}(2, \mathbf{C})$ defined by $\theta(e_\alpha) = e$, $\theta(f_\alpha) = f$, $\theta(h_\alpha) = h$ is a Lie algebra isomorphism.

Hint: With the notation used in the statement of the lemma, one can take $e_\alpha = x$ and $f_\alpha = \lambda y$ for a suitable choice of $\lambda \in \mathbf{C}$.

10.5 Root Strings and Eigenvalues

We can use the Killing form to define an isomorphism between H and $H^\star$ as follows. Given $h \in H$, let θ_h denote the map $\theta_h \in H^\star$ defined by

$$\theta_h(k) = \kappa(h, k) \text{ for all } k \in H.$$

By Lemma 10.1(iii) the Killing form is non-degenerate on restriction to H, so the map $h \mapsto \theta_h$ is an isomorphism between H and $H^\star$. (If you did Exercise 9.10 then you will have seen this before; the proof is outlined in Appendix A.) In particular, associated to each root $\alpha \in \Phi$ there is a unique element $t_\alpha \in H$ such that

$$\kappa(t_\alpha, k) = \alpha(k) \text{ for all } k \in H.$$

One very useful property of this correspondence is the following lemma.

Lemma 10.6

Let $\alpha \in \Phi$. If $x \in L_\alpha$ and $y \in L_{-\alpha}$, then $[x, y] = \kappa(x, y)t_\alpha$. In particular, $h_\alpha = [e_\alpha, f_\alpha] \in \mathrm{Span}\{t_\alpha\}$.

Proof

For $h \in H$, we have

$$\kappa(h, [x, y]) = \kappa([h, x], y) = \alpha(h)\kappa(x, y) = \kappa(t_\alpha, h)\kappa(x, y).$$

Now we view $\kappa(x, y)$ as a scalar and rewrite the right-hand side to get

$$\kappa(h, [x, y]) = \kappa(h, \kappa(x, y)t_\alpha).$$

This shows that $[x, y] - \kappa(x, y)t_\alpha$ is perpendicular to all $h \in H$, and hence it is zero as κ restricted to H is non-degenerate. $\qquad\square$

We are now in a position to apply the results of Chapter 8 on the representation theory of $\mathsf{sl}(2, \mathbf{C})$. Let α be a root. We may regard L as an $\mathsf{sl}(\alpha)$-module via restriction of the adjoint representation. Thus, if $a \in \mathsf{sl}(\alpha)$ and $y \in L$, then the action is given by

$$a \cdot y = (\mathrm{ad}\, a)y = [a, y].$$

Note that the $\mathsf{sl}(\alpha)$-submodules of L are precisely the vector subspaces M of L such that $[s, m] \in M$ for all $s \in \mathsf{sl}(\alpha)$ and $m \in M$. Of course, it is enough to check this when s is one of the standard basis elements $h_\alpha, e_\alpha, f_\alpha$. We shall also need the following lemma.

Lemma 10.7

If M is an $\mathsf{sl}(\alpha)$-submodule of L, then the eigenvalues of h_α acting on M are integers.

Proof

By Weyl's Theorem, M may be decomposed into a direct sum of irreducible $\mathsf{sl}(\alpha)$-modules; for irreducible $\mathsf{sl}(2, \mathbf{C})$-modules, the result follows from the classification of Chapter 8. $\square$

Example 10.8

(1) If you did Exercise 8.3, then you will have seen how $\mathsf{sl}(3, \mathbf{C})$ decomposes as an $\mathsf{sl}(\alpha)$-module where $\alpha = \varepsilon_1 - \varepsilon_2$ is a root of the Cartan subalgebra of $\mathsf{sl}(3, \mathbf{C})$ consisting of all diagonal matrices.

(2) Let $U = H + \mathsf{sl}(\alpha)$. Let $K = \ker \alpha \subseteq H$. By the rank-nullity formula, $\dim K = \dim H - 1$. (We know that $\dim \operatorname{im} \alpha = 1$ as $\alpha(h_\alpha) \neq 0$.) As H is abelian, $[h_\alpha, x] = 0$ for all $x \in K$. Moreover, if $x \in K$, then

$$[e_\alpha, x] = -[x, e_\alpha] = -\alpha(x)e_\alpha = 0$$

and similarly $[f_\alpha, x] = 0$. Thus every element of $\mathsf{sl}(\alpha)$ acts trivially on K. It follows that $U = K \oplus \mathsf{sl}(\alpha)$ is a decomposition of U into $\mathsf{sl}(\alpha)$-modules. By Exercise 8.2(iii), the adjoint representation of $\mathsf{sl}(\alpha)$ is isomorphic to V_2, so U is isomorphic to the direct sum of $\dim H - 1$ copies of the trivial representation, V_0, and one copy of the adjoint representation, V_2.

(3) If $\beta \in \Phi$ or $\beta = 0$, let

$$M := \bigoplus_c L_{\beta + c\alpha},$$

where the sum is over all $c \in \mathbf{C}$ such that $\beta + c\alpha \in \Phi$. It follows from Lemma 10.1(i) that M is an $\mathsf{sl}(\alpha)$-submodule of L. This module is said to be the α-*root string through* β. Analysing these modules will give the main results of this section.

Proposition 10.9

Let $\alpha \in \Phi$. The root spaces $L_{\pm \alpha}$ are 1-dimensional. Moreover, the only multiples of α which lie in Φ are $\pm \alpha$.

Proof

If $c\alpha$ is a root, then h_α takes $c\alpha(h_\alpha) = 2c$ as an eigenvalue. As the eigenvalues of h_α are integral, either $c \in \mathbf{Z}$ or $c \in \mathbf{Z} + \frac{1}{2}$. To rule out the unwanted values for c, we consider the root string module

$$M := H \oplus \bigoplus_{c\alpha \in \Phi} L_{c\alpha}.$$

Let $K = \ker \alpha \subseteq H$. By Example 10.8(2) above, $K \oplus \mathsf{sl}(\alpha)$ is an $\mathsf{sl}(\alpha)$-submodule of M. By Weyl's Theorem, modules for $\mathsf{sl}(\alpha)$ are completely reducible, so we may write

$$M = K \oplus \mathsf{sl}(\alpha) \oplus W,$$

where W is a complementary submodule.

If either of the conclusions of the proposition are false, then W is non-zero. Let $V \cong V_s$ be an irreducible submodule of W. If s is even, then it follows from the classification of Chapter 8 that V contains an h_α-eigenvector with eigenvalue 0. Call this eigenvector v. The zero-eigenspace of h_α on M is H, which is contained in $K \oplus \mathsf{sl}(\alpha)$. Hence $v \in (K \oplus \mathsf{sl}(\alpha)) \cap V = 0$, which is a contradiction.

Before considering the case where s is odd, we pursue another consequence of this argument. Suppose that $2\alpha \in \Phi$. Then h_α has $2\alpha(h_\alpha) = 4$ as an eigenvalue. As the eigenvalues of h_α on $K \oplus \mathsf{sl}(\alpha)$ are 0 and ± 2, the only way this could happen is if W contains an irreducible submodule V_s with s even, which we just saw is impossible.

Now suppose that s is odd. Then V must contain an h_α-eigenvector with eigenvalue 1. As $\alpha(h_\alpha) = 2$, this implies that $\frac{1}{2}\alpha$ is a root of L. But then both $\frac{1}{2}\alpha$ and α are roots of L, which contradicts the previous paragraph. $\quad\square$

Proposition 10.10

Suppose that $\alpha, \beta \in \Phi$ and $\beta \neq \pm\alpha$.

(i) $\beta(h_\alpha) \in \mathbf{Z}$.

(ii) There are integers $r, q \geq 0$ such that if $k \in \mathbf{Z}$, then $\beta + k\alpha \in \Phi$ if and only if $-r \leq k \leq q$. Moreover, $r - q = \beta(h_\alpha)$.

(iii) If $\alpha + \beta \in \Phi$, then $[e_\alpha, e_\beta]$ is a non-zero scalar multiple of $e_{\alpha+\beta}$.

(iv) $\beta - \beta(h_\alpha)\alpha \in \Phi$.

Proof

Let $M := \bigoplus_k L_{\beta+k\alpha}$ be the root string of α through β. To prove (i), we note that $\beta(h_\alpha)$ is the eigenvalue of h_α acting on L_β, and so it lies in $\mathbf{Z}$.

We know from the previous proposition that $\dim L_{\beta+k\alpha} = 1$ whenever $\beta+k\alpha$ is a root, so the eigenspaces of $\operatorname{ad} h_\alpha$ on M are all 1-dimensional and, since $(\beta + k\alpha)h_\alpha = \beta(h_\alpha) + 2k$, the eigenvalues of $\operatorname{ad} h_\alpha$ on M are either all even or all odd. It now follows from Chapter 8 that M is an irreducible $\mathsf{sl}(\alpha)$-module. Suppose that $M \cong V_d$. On V_d, the element h_α acts diagonally with eigenvalues

$$\{d, d-2, \ldots, -d\},$$

whereas on M, h_α acts diagonally with eigenvalues

$$\{\beta(h_\alpha) + 2k : \beta + k\alpha \in \Phi\}.$$

Equating these sets shows that if we define r and q by $d = \beta(h_\alpha) + 2q$ and $-d = \beta(h_\alpha) - 2r$, then (ii) will hold.

Suppose $v \in L_\beta$, so v belongs to the h_α-eigenspace where h_α acts as $\beta(h_\alpha)$. If $(\operatorname{ad} e_\alpha)e_\beta = 0$, then e_β is a highest-weight vector in the irreducible representation $M \cong V_d$, with highest weight $\beta(h_\alpha)$. If $\alpha + \beta$ is a root, then h_α acts on the associated root space as $(\alpha + \beta)h_\alpha = \beta(h_\alpha) + 2$. Therefore e_β is *not* in the highest weight space of the irreducible representation M, and so $(\operatorname{ad} e_\alpha)e_\beta \neq 0$. This proves (iii).

Finally, (iv) follows from part (ii) as

$$\beta - \beta(h_\alpha)\alpha = \beta - (r - q)\alpha$$

and $-r \leq -r + q \leq q$. $\square$

We now have a good idea about the structure constant of L (with respect to a basis given by the root space decomposition). The action of H on the root spaces of L is determined by the roots. Part (iii) of the previous proposition shows that (up to scalar factors) the set of roots also determines the brackets $[e_\alpha, e_\beta]$ for roots $\alpha \neq \pm\beta$. Lastly, by construction, $[e_\alpha, e_{-\alpha}]$ is in the span of $[e_\alpha, f_\alpha] = h_\alpha$. The reader keen to see a complete answer should read about Chevalley's Theorem in §15.3.

10.6 Cartan Subalgebras as Inner-Product Spaces

We conclude this chapter by showing that the roots of L all lie in a real vector subspace of $H^\star$ and that the Killing form induces an inner product on the space. This will enable us to bring some elementary geometric ideas to bear on the classification problem.

The two propositions in the previous section show that the set Φ of roots cannot be too big: For example, we saw that if $\alpha \in \Phi$ then the only multiples of $\alpha \in \Phi$ are $\pm\alpha$. On the other hand, there must be roots, as otherwise the root space decomposition would imply that $L = H$ was abelian. What more can be said?

Lemma 10.11

(i) If $h \in H$ and $h \neq 0$, then there exists a root $\alpha \in \Phi$ such that $\alpha(h) \neq 0$.

(ii) The set Φ of roots spans $H^\star$.

Proof

Suppose that $\alpha(h) = 0$ for all roots α. Then we have $[h, x] = \alpha(h)x = 0$ for all $x \in L_\alpha$ and for all roots α. Since H is abelian, it follows from the root space decomposition that $h \in Z(L)$, which is zero as L is semisimple.

In a sense, (ii) is just a reformulation of (i) in the language of linear algebra. Let $W \subseteq H^\star$ denote the span of Φ. Suppose that W is a proper subspace of $H^\star$. Then the annihilator of W in H,

$$W^\circ = \{h \in H : \theta(h) = 0 \text{ for all } \theta \in W\},$$

has dimension $\dim H - \dim W \neq 0$. (See Appendix A.) Therefore there is some non-zero $h \in H$ such that $\theta(h) = 0$ for all $\theta \in W$, so in particular $\alpha(h) = 0$ for all $\alpha \in \Phi$, in contradiction to part (i). $\square$

In the previous section, we found that the elements t_α and h_α spanned the same 1-dimensional subspace of L. More precisely, we have the following.

Lemma 10.12

For each $\alpha \in \Phi$, we have

(i) $t_\alpha = \dfrac{h_\alpha}{\kappa(e_\alpha, f_\alpha)}$ and $h_\alpha = \dfrac{2t_\alpha}{\kappa(t_\alpha, t_\alpha)}$;

(ii) $\kappa(t_\alpha, t_\alpha)\kappa(h_\alpha, h_\alpha) = 4$.

Proof

The expression for t_α follows from Lemma 10.6 applied with $x = e_\alpha$ and $y = f_\alpha$.

As $\alpha(h_\alpha) = 2$, we have

$$2 = \kappa(t_\alpha, h_\alpha) = \kappa(t_\alpha, \kappa(e_\alpha, f_\alpha)t_\alpha),$$

which implies that $\kappa(e_\alpha, f_\alpha)\kappa(t_\alpha, t_\alpha) = 2$. Now substitute for $\kappa(e_\alpha, f_\alpha)$ to get the second expression. Finally,

$$\kappa(h_\alpha, h_\alpha) = \kappa\left(\frac{2t_\alpha}{\kappa(t_\alpha, t_\alpha)}, \frac{2t_\alpha}{\kappa(t_\alpha, t_\alpha)}\right) = \frac{4}{\kappa(t_\alpha, t_\alpha)}$$

gives (ii).   □

Corollary 10.13

If α and β are roots, then $\kappa(h_\alpha, h_\beta) \in \mathbf{Z}$ and $\kappa(t_\alpha, t_\beta) \in \mathbf{Q}$.

Proof

Using the root space decomposition to compute $\mathrm{tr}(\mathrm{ad}\, h_\alpha \circ \mathrm{ad}\, h_\beta)$, we get

$$\kappa(h_\alpha, h_\beta) = \sum_{\gamma \in \Phi} \gamma(h_\alpha)\gamma(h_\beta).$$

Since the eigenvalues of h_α and h_β are integers, this shows that $\kappa(h_\alpha, h_\beta) \in \mathbf{Z}$. We now use the previous lemma to get

$$\kappa(t_\alpha, t_\beta) = \kappa\left(\frac{\kappa(t_\alpha, t_\alpha)h_\alpha}{2}, \frac{\kappa(t_\beta, t_\beta)h_\beta}{2}\right)$$

$$= \frac{\kappa(t_\alpha, t_\alpha)\kappa(t_\beta, t_\beta)}{4}\kappa(h_\alpha, h_\beta) \in \mathbf{Q}. \qquad \square$$

We can translate the Killing form on H to obtain a non-degenerate symmetric bilinear form on H^*, denoted $(-, -)$. This form may be defined by

$$(\theta, \varphi) = \kappa(t_\theta, t_\varphi),$$

where t_θ and t_φ are the elements of H corresponding to θ and φ under the isomorphism $H \equiv H^\star$ induced by κ. In particular, if α and β are roots, then

$$(\alpha, \beta) = \kappa(t_\alpha, t_\beta) \in \mathbf{Q}.$$

Exercise 10.4

Show that $\beta(h_\alpha) = \frac{2(\beta, \alpha)}{(\alpha, \alpha)}$.

We saw in Lemma 10.11 that the roots of L span $H^\star$, so $H^\star$ has a vector space basis consisting of roots, say $\{\alpha_1, \alpha_2, \ldots, \alpha_\ell\}$. We can now prove that something stronger is true as follows.

Lemma 10.14

If β is a root, then β is a linear combination of the α_i with coefficients in $\mathbf{Q}$.

Proof

Certainly we may write $\beta = \sum_{i=1}^{\ell} c_i \alpha_i$ with coefficients $c_i \in \mathbf{C}$. For each j with $1 \leq j \leq \ell$, we have

$$(\beta, \alpha_j) = \sum_{i=1}^{\ell} (\alpha_i, \alpha_j) c_i.$$

We can write these equations in matrix form as

$$\begin{pmatrix} (\beta, \alpha_1) \\ \vdots \\ (\beta, \alpha_\ell) \end{pmatrix} = \begin{pmatrix} (\alpha_1, \alpha_1) & \cdots & (\alpha_\ell, \alpha_1) \\ \vdots & \ddots & \vdots \\ (\alpha_1, \alpha_\ell) & \cdots & (\alpha_\ell, \alpha_\ell) \end{pmatrix} \begin{pmatrix} c_1 \\ \vdots \\ c_\ell \end{pmatrix}.$$

The matrix is the matrix of the non-degenerate bilinear form $(-, -)$ with respect to the chosen basis of roots, and so it is invertible (see Appendix A). Moreover, we have seen that its entries are rational numbers, so it has an inverse with entries in $\mathbf{Q}$. Since also $(\beta, \alpha_j) \in \mathbf{Q}$, the coefficients c_i are rational. $\qquad \square$

By this lemma, the *real* subspace of $H^\star$ spanned by the roots $\alpha_1, \ldots, \alpha_\ell$ contains all the roots of $\varPhi$ and so does not depend on our particular choice of basis. Let E denote this subspace.

Proposition 10.15

The form $(-, -)$ is a real-valued inner product on E.

Proof

Since $(-, -)$ is a symmetric bilinear form, we only need to check that the restriction of $(-, -)$ to E is positive definite. Let $\theta \in E$ correspond to $t_\theta \in H$. Using the root space decomposition and the fact that $(\operatorname{ad} t_\theta) e_\beta = \beta(t_\theta) e_\beta$, we get

$$(\theta, \theta) = \kappa(t_\theta, t_\theta) = \sum_{\beta \in \varPhi} \beta(t_\theta)^2 = \sum_{\beta \in \varPhi} \kappa(t_\beta, t_\theta)^2 = \sum_{\beta \in \varPhi} (\beta, \theta)^2.$$

As (β, θ) is real, the right-hand side is real and non-negative. Moreover, if $(\theta, \theta) = 0$, then $\beta(t_\theta) = 0$ for all roots β, so by Lemma 10.11(i), $\theta = 0$. $\qquad \square$

EXERCISES

10.5.   Suppose that L is a complex semisimple Lie algebra with Cartan subalgebra H and root system Φ. Use the results of this chapter to prove that

$$\dim L = \dim H + |\Phi|.$$

Hence show that there are no semisimple Lie algebras of dimensions 4, 5, or 7.

10.6.†   Let $L = \mathsf{sl}(3, \mathbf{C})$. With the same notation as in the introduction, let $\alpha := \varepsilon_1 - \varepsilon_2$ and $\beta := \varepsilon_2 - \varepsilon_3$. Show that the set of roots is

$$\Phi = \{\pm\alpha, \pm\beta \pm (\alpha + \beta)\}.$$

Show that the angle between the roots α and β is $2\pi/3$ and verify some of the results in §10.5 and §10.6 for $\mathsf{sl}(3, \mathbf{C})$.

10.7.   Suppose L is semisimple of dimension 6. Let H be a Cartan subalgebra of L and let Φ be the associated set of roots.

(i)   Show that $\dim H = 2$ and that if $\alpha, \beta \in \Phi$ span $H^\star$, then $\Phi = \{\pm\alpha, \pm\beta\}$.

(ii)   Hence show that

$$[L_\alpha, L_{\pm\beta}] = 0 \text{ and } [L_{\pm\beta}, [L_\alpha, L_{-\alpha}]] = 0$$

and deduce that the subalgebra $L_\alpha \oplus L_{-\alpha} \oplus [L_\alpha, L_{-\alpha}]$ is an ideal of L. Show that L is isomorphic to the direct sum of two copies of $\mathsf{sl}(2, \mathbf{C})$.

10.8.   Show that the set of diagonal matrices in $\mathsf{so}(4, \mathbf{C})$ (as defined in Chapter 4) forms a Cartan subalgebra of $\mathsf{so}(4, \mathbf{C})$ and determine the corresponding root space decomposition. Hence show that $\mathsf{so}(4, \mathbf{C}) \cong \mathsf{sl}(2, \mathbf{C}) \oplus \mathsf{sl}(2, \mathbf{C})$. (The reader will probably now be able to guess the reason for choosing non-obvious bilinear forms in the definition of the symplectic and orthogonal Lie algebras.)

10.9.   Let L be a semisimple Lie algebra with Cartan subalgebra H. Use the root space decomposition to show that $N_L(H) = H$. (The notation $N_L(H)$ is defined in Exercise 5.6.)

10.10.   In Lemma 10.5, we defined for each $\alpha \in \Phi$ a subalgebra of L isomorphic to $\mathsf{sl}(2, \mathbf{C})$. In the main step in the proof, we showed that if $x \in L_\alpha$ and $y \in L_{-\alpha}$, and $h = [x, y] \neq 0$, then $\alpha(h) \neq 0$. Here is an alternative proof of this using root string modules.

Suppose that $\alpha(h) = 0$. Let $\beta \in \Phi$ be any root. Let M be the α root string module through β,

$$M = \bigoplus_c L_{\beta + c\alpha}.$$

By considering the trace of h on M, show that $\beta(h) = 0$ and hence get a contradiction.

10.11.   Let L be a complex semisimple Lie algebra with Cartan subalgebra H and root space Φ. Let $\alpha \in \Phi$ and let $\mathsf{sl}(\alpha) = \mathrm{Span}\{e_\alpha, f_\alpha, h_\alpha\}$ be the corresponding subalgebra constructed in §10.4. Show that this subalgebra is unique up to

(1) scaling basis elements as $ce_\alpha, c^{-1}f_\alpha, h_\alpha$ for non-zero $c \in \mathbf{C}$; and

(2) swapping e_α and f_α and then replacing h_α with $-h_\alpha$.

10.12.†   Let L be a semisimple Lie algebra and let Φ be its set of roots. Let $\alpha \in \Phi$ and let

$$N := \mathrm{Span}\{f_\alpha\} \oplus \mathrm{Span}\{h_\alpha\} \oplus L_\alpha \oplus L_{2\alpha} \oplus \ldots .$$

Show that N is an $\mathsf{sl}(\alpha)$-submodule of L. By considering the trace of $h_\alpha : N \to N$, give an alternative proof of Proposition 10.9.

11
Root Systems

The essential properties of the roots of complex semisimple Lie algebras may be captured in the idea of an abstract "root system". In this chapter, we shall develop the basic theory of root systems. Our eventual aim, achieved in Chapters 13 and 14, will be to use root systems to classify the complex semisimple Lie algebras.

Root systems have since been discovered to be important in many other areas of mathematics, so while this is probably your first encounter with root systems, it may well not be your last! In MathSciNet, the main database for research papers in mathematics, there are, at the time of writing, 297 papers whose title contains the words "root system", and many thousands more in which root systems are mentioned in the text.

11.1 Definition of Root Systems

Let E be a finite-dimensional real vector space endowed with an inner product written $(-, -)$. Given a non-zero vector $v \in E$, let s_v be the reflection in the hyperplane normal to v. Thus s_v sends v to $-v$ and fixes all elements y such that $(y, v) = 0$. As an easy exercise, the reader may check that

$$s_v(x) = x - \frac{2(x, v)}{(v, v)} v \quad \text{for all } x \in E$$

and that s_v preserves the inner product, that is,

$$(s_v(x), s_v(y)) = (x, y) \quad \text{for all } x, y \in E.$$

As it is a very useful convention, we shall write

$$\langle x, v \rangle := \frac{2(x, v)}{(v, v)},$$

noting that the symbol $\langle x, v \rangle$ is only linear with respect to its first variable, x. With this notation, we can now define root systems.

Definition 11.1

A subset R of a real inner-product space E is a *root system* if it satisfies the following axioms.

(R1) R is finite, it spans E, and it does not contain 0.

(R2) If $\alpha \in R$, then the only scalar multiples of α in R are $\pm\alpha$.

(R3) If $\alpha \in R$, then the reflection s_α permutes the elements of R.

(R4) If $\alpha, \beta \in R$, then $\langle \beta, \alpha \rangle \in \mathbf{Z}$.

The elements of R are called *roots*.

Example 11.2

The root space decomposition gives our main example. Let L be a complex semisimple Lie algebra, and suppose that Φ is the set of roots of L with respect to some fixed Cartan subalgebra H. Let E denote the real span of Φ. By Proposition 10.15, the symmetric bilinear form on E induced by the Killing form $(-, -)$ is an inner product.

We can use the results of §10.5 and §10.6 to show that Φ is a root system in E. By definition, $0 \notin \Phi$ and, as we observed early on, Φ is finite. We showed that (R2) holds in Proposition 10.9. To show that (R3) holds, we note that if $\alpha, \beta \in \Phi$ then

$$s_\alpha(\beta) = \beta - \frac{2(\beta, \alpha)}{(\alpha, \alpha)}\alpha = \beta - \beta(h_\alpha)\alpha,$$

which lies in Φ by Proposition 10.10. To get the second equality above, we used the identity of Exercise 10.4, which may be proved as follows:

$$\beta(h_\alpha) = \kappa(t_\beta, h_\alpha) = \kappa\left(t_\beta, \frac{2t_\alpha}{(t_\alpha, t_\alpha)}\right) = \frac{2(\beta, \alpha)}{(\alpha, \alpha)} = \langle \beta, \alpha \rangle.$$

As the eigenvalues of h_α are integers, this identity also establishes (R4).

Exercise 11.1

We work in $\mathbf{R}^{\ell+1}$, with the Euclidean inner product. Let ε_i be the vector in $\mathbf{R}^{\ell+1}$ with i-th entry 1 and all other entries zero. Define

$$R := \{\pm(\varepsilon_i - \varepsilon_j) : 1 \le i < j \le \ell + 1\}$$

and let $E = \operatorname{Span} R = \{\sum \alpha_i \varepsilon_i : \sum \alpha_i = 0\}$. Show that R is a root system in E.

Remark 11.3

We shall see that our axioms isolate all the essential properties of roots of Lie algebras. For this reason, there is no need in this chapter to keep the full body of theory we have developed in mind — doing so would burden us with extraneous notions while needlessly reducing the applicability of our arguments. In any case, we shall see later that every root system *is* the set of roots of a complex semisimple Lie algebra, so our problem is no more general than is necessary: *"It is the mark of the educated mind to use for each subject the degree of exactness which it admits"* (Aristotle).

11.2 First Steps in the Classification

The following lemma gives the first indication that the axioms for root systems are quite restrictive.

Lemma 11.4 (Finiteness Lemma)

Suppose that R is a root system in the real inner-product space E. Let $\alpha, \beta \in R$ with $\beta \ne \pm\alpha$. Then

$$\langle \alpha, \beta \rangle \langle \beta, \alpha \rangle \in \{0, 1, 2, 3\}.$$

Proof

Thanks to (R4), the product in question is an integer: We must establish the bounds. For any non-zero $v, w \in E$, the angle θ between v and w is such that $(v, w)^2 = (v, v)(w, w) \cos^2 \theta$. This gives

$$\langle \alpha, \beta \rangle \langle \beta, \alpha \rangle = 4 \cos^2 \theta \le 4.$$

Suppose we have $\cos^2 \theta = 1$. Then θ is an integer multiple of π and so α and β are linearly dependent, contrary to our assumption. $\qquad\square$

We now use this lemma to show that there are only a few possibilities for the integers $\langle \alpha, \beta \rangle$. Take two roots α, β in a root system R with $\alpha \neq \pm\beta$. We may choose the labelling so that $(\beta, \beta) \geq (\alpha, \alpha)$ and hence

$$|\langle \beta, \alpha \rangle| = \frac{2\,|(\beta, \alpha)|}{(\alpha, \alpha)} \geq \frac{2\,|(\alpha, \beta)|}{(\beta, \beta)} = |\langle \alpha, \beta \rangle|\,.$$

By the Finiteness Lemma, the possibilities are:

$\langle \alpha, \beta \rangle$	$\langle \beta, \alpha \rangle$	θ	$\dfrac{(\beta, \beta)}{(\alpha, \alpha)}$
0	0	$\pi/2$	undetermined
1	1	$\pi/3$	1
-1	-1	$2\pi/3$	1
1	2	$\pi/4$	2
-1	-2	$3\pi/4$	2
1	3	$\pi/6$	3
-1	-3	$5\pi/6$	3

Given roots α and β, we would like to know when their sum and difference lie in R. Our table gives some information about this question.

Proposition 11.5

Let $\alpha, \beta \in R$.

(a) If the angle between α and β is strictly obtuse, then $\alpha + \beta \in R$.

(b) If the angle between α and β is strictly acute and $(\beta, \beta) \geq (\alpha, \alpha)$, then $\alpha - \beta \in R$.

Proof

In either case, we may assume that $(\beta, \beta) \geq (\alpha, \alpha)$. By (R3), we know that $s_\beta(\alpha) = \alpha - \langle \alpha, \beta \rangle \beta$ lies in R. The table shows that if θ is strictly acute, then $\langle \alpha, \beta \rangle = 1$, and if θ is strictly obtuse, then $\langle \alpha, \beta \rangle = -1$.   □

Example 11.6

Let $E = \mathbf{R}^2$ with the Euclidean inner product. We shall find all root systems R contained in E. Take a root α of the shortest possible length. Since R spans E, it must contain some root $\beta \neq \pm\alpha$. By considering $-\beta$ if necessary, we may assume that β makes an obtuse angle with α. Moreover, we may assume that this angle, say θ, is as large as possible.

(a) Suppose that $\theta = 2\pi/3$. Using Proposition 11.5, we find that R contains the six roots shown below.

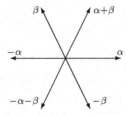

One can check that this set is closed under the action of the reflections $s_\alpha, s_\beta, s_{\alpha+\beta}$. As $s_{-\alpha} = s_\alpha$, and so on, this is sufficient to verify (R3). We have therefore found a root system in E. This root system is said to have *type* A_2. (The 2 refers to the dimension of the underlying space.)

(b) Suppose that $\theta = 3\pi/4$. Proposition 11.5 shows that $\alpha + \beta$ is a root, and applying s_α to β shows that $2\alpha + \beta$ is a root, so R must contain

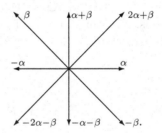

This root system is said to have *type* B_2. A further root would make an angle of at most $\pi/8$ with one of the existing roots, so this must be all of R.

(c) Suppose that $\theta = 5\pi/6$. We leave it to the reader to show that R must be

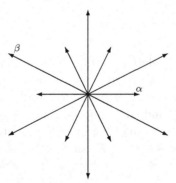

and to determine the correct labels for the remaining roots. This root system is said to have *type* G_2.

(d) Suppose that β is perpendicular to α. This gives us the root system of *type $A_1 \times A_1$.*

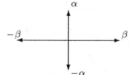

Here, as $(\alpha, \beta) = 0$, the reflection s_α fixes the roots $\pm\beta$ lying in the space perpendicular to α, so there is no interaction between the roots $\pm\alpha$ and $\pm\beta$. In particular, knowing the length of α tells us nothing about the length of β. These considerations suggest the following definition.

Definition 11.7

The root system R is *irreducible* if R cannot be expressed as a disjoint union of two non-empty subsets $R_1 \cup R_2$ such that $(\alpha, \beta) = 0$ for $\alpha \in R_1$ and $\beta \in R_2$.

Note that if such a decomposition exists, then R_1 and R_2 are root systems in their respective spans. The next lemma tells us that it will be enough to classify the irreducible root systems.

Lemma 11.8

Let R be a root system in the real vector space E. We may write R as a disjoint union
$$R = R_1 \cup R_2 \cup \ldots \cup R_k,$$
where each R_i is an irreducible root system in the space E_i spanned by R_i, and E is a direct sum of the orthogonal subspaces $E_1, \ldots, E_k$.

Proof

Define an equivalence relation $\sim$ on R by letting $\alpha \sim \beta$ if there exist $\gamma_1, \gamma_2, \ldots, \gamma_s$ in R with $\alpha = \gamma_1$ and $\beta = \gamma_s$ such that $(\gamma_i, \gamma_{i+1}) \neq 0$ for $1 \leq i < s$. Let the R_i be the equivalence classes for this relation. It is clear that they satisfy axioms (R1), (R2), and (R4); you are asked to check (R3) in the following exercise. That each R_i is irreducible follows immediately from the construction.

As every root appears in some E_i, the sum of the E_i spans E. Suppose that $v_1 + \ldots + v_k = 0$, where $v_i \in E_i$. Taking inner products with v_j, we get

$$0 = (v_1, v_j) + \ldots + (v_j, v_j) + \ldots + (v_k, v_j) = (v_j, v_j)$$

so each $v_j = 0$. Hence $E = E_1 \oplus \ldots \oplus E_k$. $\square$

Exercise 11.2

Show that if $(\alpha, \beta) \neq 0$, then $(\alpha, s_\alpha(\beta)) \neq 0$. Deduce that the equivalence classes defined in the proof of the lemma satisfy (R3).

11.3 Bases for Root Systems

Let R be a root system in the real inner-product space E. Because R spans E, any maximal linearly independent subset of R is a vector space basis for R. Proposition 11.5 suggests that it might be convenient if we could find such a subset where every pair of elements made an obtuse angle. In fact, we can ask for something stronger, as in the following.

Definition 11.9

A subset B of R is a *base* for the root system R if

(B1) B is a vector space basis for E, and

(B2) every $\beta \in R$ can be written as $\beta = \sum_{\alpha \in B} k_\alpha \alpha$ with $k_\alpha \in \mathbf{Z}$, where all the non-zero coefficients k_α have the same sign.

Exercise 11.3

Show that if B is a base for a root system, then the angle between any two distinct elements of B is obtuse (that is, at least $\pi/2$).

We say that a root $\beta \in R$ is *positive with respect to B* if the coefficients given in (B2) are positive, and otherwise it is *negative with respect to B*.

Exercise 11.4

Let $R = \{\pm(\varepsilon_i - \varepsilon_j) : 1 \le i < j \le \ell + 1\}$ be the root system in Exercise 11.1. Let $\alpha_i = \varepsilon_i - \varepsilon_{i+1}$ for $1 \le i \le \ell$. Show that $B = \{\alpha_1, \ldots, \alpha_\ell\}$ is a base for R and find the positive roots.

A natural way to label the elements of R as positive or negative is to fix a hyperplane of codimension 1 in E which does not contain any element of R and then to label the roots of one side of the hyperplane as positive and those on the other side as negative. Suppose that R has a base B compatible with this labelling. Then the elements of B must lie on the positive side of the hyperplane. For example, the diagram below shows a possible base for the root system in Example 11.6(b).

$\{\alpha, \beta\}$ is a base of the root system of type B_2

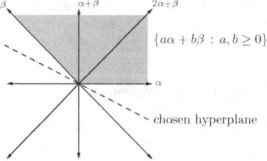

Note that the roots in the base are those nearest to the hyperplane. This observation motivates the proof of our next theorem.

Theorem 11.10

Every root system has a base.

Proof

Let R be a root system in the real inner-product space E. We may assume that E has dimension at least 2, as the case $\dim E = 1$ is obvious. We may choose a vector $z \in E$ which does not lie in the perpendicular space of any of the roots. Such a vector must exist, as E has dimension at least 2, so it is not the union of finitely many hyperplanes (see Exercise 11.8 or Exercise 11.12).

Let R^+ be the set of $\alpha \in R$ which lie on the positive side of z, that is, those α for which $(z, \alpha) > 0$. Let

$$B := \{\alpha \in R^+ : \alpha \text{ is not the sum of two elements in } R^+\}.$$

We claim that B is a base for R.

We first show that (B2) holds. If $\beta \in R$, then either $\beta \in R^+$ or $-\beta \in R^+$, so it is sufficient to prove that every $\beta \in R^+$ can be expressed as $\beta = \sum_{\alpha \in B} k_\alpha \alpha$ for some $k_\alpha \in \mathbf{Z}$ with each $k_\alpha \geq 0$. If this fails, then we may pick, from the elements of R^+ that are not of this form, an element $\beta \in R^+$ such that the inner product (z, β) is as small as possible. As $\beta \notin B$, there exist $\beta_1, \beta_2 \in R^+$ such that $\beta = \beta_1 + \beta_2$. By linearity,

$$(z, \beta) = (z, \beta_1) + (z, \beta_2)$$

is the sum of two positive numbers, and therefore $0 < (z, \beta_i) < (z, \beta)$ for

$i = 1, 2$. Now at least one of β_1, β_2 cannot be expressed as a positive integral linear combination of the elements of B; this contradicts the choice of β.

It remains to show that B is linearly independent. First note that if α and β are distinct elements of B, then by Exercise 11.3 the angle between them must be obtuse. Suppose that $\sum_{\alpha \in B} r_\alpha \alpha = 0$, where $r_\alpha \in \mathbf{R}$. Collecting all the terms with positive coefficients to one side gives an element

$$x := \sum_{\alpha : r_\alpha > 0} r_\alpha \alpha = \sum_{\beta : r_\beta < 0} (-r_\beta) \beta.$$

Hence

$$(x, x) = \sum_{\substack{\alpha : r_\alpha > 0 \\ \beta : r_\beta < 0}} r_\alpha (-r_\beta)(\alpha, \beta) \le 0$$

and so $x = 0$. Therefore

$$0 = (x, z) = \sum_{\alpha : r_\alpha > 0} r_\alpha (\alpha, z),$$

where each $(\alpha, z) > 0$ as $\alpha \in R^+$, so we must have $r_\alpha = 0$ for all α, and similarly $r_\beta = 0$ for all β. □

Let R^+ denote the set of all positive roots in a root system R with respect to a base B, and let R^- be the set of all negative roots. Then $R = R^+ \cup R^-$, a disjoint union. The set B is contained in R^+; the elements of B are called *simple roots*. The reflections s_α for $\alpha \in B$ are known as *simple reflections*.

Remark 11.11

A root system R will usually have many possible bases. For example, if B is a base then so is $\{-\alpha : \alpha \in B\}$. In particular, the terms "positive" and "negative" roots are always taken with reference to a fixed base B.

Exercise 11.5

Let R be a root system with a base B. Take any $\gamma \in R$. Show that the set $\{s_\gamma(\alpha) : \alpha \in B\}$ is also a base of R.

11.3.1 The Weyl Group of a Root System

For each root $\alpha \in R$, we have defined a reflection s_α which acts as an invertible linear map on E. We may therefore consider the group of invertible linear transformations of E generated by the reflections s_α for $\alpha \in R$. This is known as the *Weyl group* of R and is denoted by W or $W(R)$.

Lemma 11.12

The Weyl group W associated to R is finite.

Proof

By axiom (R3), the elements of W permute R, so there is a group homomorphism from W into the group of all permutations of R, which is a finite group because R is finite. We claim that this homomorphism is injective, and so W is finite.

Suppose that $g \in W$ is in the kernel of this homomorphism. Then, by definition, g fixes all the roots in R. But E is spanned by the roots, so g fixes all elements in a basis of E, and so g must be the identity map. $\qquad\square$

11.3.2 Recovering the Roots

Suppose that we are given a base B for a root system R. We shall show that this alone gives us enough information to recover R. To do this, we use the Weyl group and prove that every root β is of the form $\beta = g(\alpha)$ for some $\alpha \in B$ and some g in the subgroup $W_0 := \langle s_\gamma : \gamma \in B \rangle$ of W. (We shall also see that $W = W_0$.) Thus, if we repeatedly apply reflections in the simple roots, we will eventually recover the full root system.

Some evidence for this statement is given by Example 11.6: in each case we started with a pair of roots $\{\alpha, \beta\}$, and knowing only the positions of α and β, we used repeated reflections to construct the unique root system containing these roots as a base.

Lemma 11.13

If $\alpha \in B$, then the reflection s_α permutes the set of positive roots other than α.

Proof

Suppose that $\beta \in R^+$ and $\beta \neq \alpha$. We know that $\beta = \sum_{\gamma \in B} k_\gamma \gamma$ for some $k_\gamma \geq 0$. Since $\beta \neq \alpha$ and $\beta \in R$, there is some $\gamma \in B$ with $k_\gamma \neq 0$ and $\gamma \neq \alpha$. We know that $s_\alpha(\beta) \in R$; and from $s_\alpha(\beta) = \beta - \langle \beta, \alpha \rangle \alpha$ we see that the coefficient of γ in $s_\alpha(\beta)$ is k_γ, which is positive. As all the non-zero coefficients in the expression of $s_\alpha(\beta)$ as a linear combination of base elements must have the same sign, this tells us that $s_\alpha(\beta)$ lies in R^+. $\qquad\square$

Proposition 11.14

Suppose that $\beta \in R$. There exists $g \in W_0$ and $\alpha \in B$ such that $\beta = g(\alpha)$.

Proof

Suppose first of all that $\beta \in R^+$ and that $\beta = \sum_{\gamma \in B} k_\gamma \gamma$ with $k_\gamma \in \mathbf{Z}$, $k_\gamma \geq 0$. We shall proceed by induction on the height of β defined by

$$\mathrm{ht}(\beta) = \sum_{\gamma \in B} k_\gamma.$$

If $\mathrm{ht}(\beta) = 1$, then $\beta \in B$, so we may take $\alpha = \beta$ and let g be the identity map. For the inductive step, suppose that $\mathrm{ht}(\beta) = n \geq 2$. By axiom (R2), at least two of the k_γ are strictly positive.

We claim that there is some $\gamma \in B$ such that $(\beta, \gamma) > 0$. If not, then $(\beta, \gamma) \leq 0$ for all $\gamma \in B$ and so

$$(\beta, \beta) = \sum_\gamma k_\gamma (\beta, \gamma) \leq 0,$$

which is a contradiction because $\beta \neq 0$. We may therefore choose some $\gamma \in B$ with $(\beta, \gamma) > 0$. Then $\langle \beta, \gamma \rangle > 0$ and so

$$\mathrm{ht}(s_\gamma(\beta)) = \mathrm{ht}(\beta) - \langle \beta, \gamma \rangle < \mathrm{ht}(\beta).$$

(We have $s_\gamma(\beta) \in R^+$ by the previous lemma.) The inductive hypothesis now implies that there exists $\alpha \in B$ and $h \in W_0$ such that $s_\gamma(\beta) = h(\alpha)$. Hence $\beta = s_\gamma(h(\alpha))$ so we may take $g = s_\gamma h$, which lies in W_0.

Now suppose that $\beta \in R^-$, so $-\beta \in R^+$. By the first part, $-\beta = g(\alpha)$ for some $g \in W_0$ and $\alpha \in B$. By linearity of g, we get $\beta = g(-\alpha) = g(s_\alpha(\alpha))$, where $g s_\alpha \in W_0$. $\qquad\square$

We end this section by proving that a base for a root system determines its full Weyl group. We need the following straightforward result.

Exercise 11.6

Suppose that α is a root and that $g \in W$. Show that $g s_\alpha g^{-1} = s_{g\alpha}$.

Lemma 11.15

We have $W_0 = W$; that is, W is generated by the s_α for $\alpha \in B$.

Proof

By definition, W is generated by the reflections s_β for $\beta \in R$, so it is sufficient to prove that $s_\beta \in W_0$ for any $\beta \in R$. By Proposition 11.14, we know that given β there is some $g \in W_0$ and $\alpha \in B$ such that $\beta = g(\alpha)$. Now the reflection s_β is equal to $g s_\alpha g^{-1} \in W_0$ by the previous exercise. $\qquad\square$

11.4 Cartan Matrices and Dynkin Diagrams

Although in general a root system can have many different bases, the following theorem shows that from a geometric point of view they are all essentially the same. As the proof of this theorem is slightly technical, we postpone it to Appendix D.

Theorem 11.16

Let R be a root system and suppose that B and B' are two bases of R, as defined in Definition 11.9. Then there exists an element g in the Weyl group $W(R)$ such that $B' = \{g(\alpha) : \alpha \in B\}$. $\qquad\square$

Let B be a base in a root system R. Fix an order on the elements of B, say $(\alpha_1, \ldots, \alpha_\ell)$. The *Cartan matrix* of R is defined to be the $\ell \times \ell$ matrix with ij-th entry $\langle \alpha_i, \alpha_j \rangle$. Since for any root β we have

$$\langle s_\beta(\alpha_i), s_\beta(\alpha_j) \rangle = \langle \alpha_i, \alpha_j \rangle,$$

it follows from Theorem 11.16 that the Cartan matrix depends only on the ordering adopted with our chosen base B and not on the base itself. Note that by (R4) the entries of the Cartan matrix are integers.

Example 11.17

(1) Let R be as in Exercise 11.4(ii). Calculation shows that the Cartan matrix with respect to the ordered base $(\alpha_1, \ldots, \alpha_\ell)$ is

$$\begin{pmatrix} 2 & -1 & 0 & \cdots & 0 & 0 \\ -1 & 2 & -1 & \cdots & 0 & 0 \\ 0 & -1 & 2 & \cdots & 0 & 0 \\ \vdots & \vdots & \vdots & \ddots & \vdots & \vdots \\ 0 & 0 & 0 & \cdots & 2 & -1 \\ 0 & 0 & 0 & \cdots & -1 & 2 \end{pmatrix}.$$

(2) Let R be the root system which we have drawn in Example 11.6(b). This has ordered base (α, β), and the corresponding Cartan matrix is

$$C = \begin{pmatrix} 2 & -1 \\ -2 & 2 \end{pmatrix}.$$

Another way to record the information given in the Cartan matrix is in a graph $\Delta = \Delta(R)$, defined as follows. The vertices of Δ are labelled by the simple roots of B. Between the vertices labelled by simple roots α and β, we draw $d_{\alpha\beta}$ lines, where

$$d_{\alpha\beta} := \langle \alpha, \beta \rangle \langle \beta, \alpha \rangle \in \{0, 1, 2, 3\}.$$

If $d_{\alpha\beta} > 1$, which happens whenever α and β have different lengths and are not orthogonal, we draw an arrow pointing from the longer root to the shorter root. This graph is called the *Dynkin diagram* of R. By Theorem 11.16, the Dynkin diagram of R is independent of the choice of base.

The graph with the same vertices and edges, but without the arrows, is known as the *Coxeter graph* of R.

Example 11.18

Using the base given in Exercise 11.4(ii), the Dynkin diagram of the root system introduced in Exercise 11.1 is

$$\underset{\alpha_1}{\circ}\!\!-\!\!-\!\!\underset{\alpha_2}{\circ}\!\!-\!\!-\!\!-\!\!-\quad\cdots\quad-\!\!-\!\!-\!\!\underset{\alpha_{\ell-2}}{\circ}\!\!-\!\!-\!\!\underset{\alpha_{\ell-1}}{\circ}.$$

The Dynkin diagram for the root system in Example 11.6(b) is

$$\underset{\beta}{\circ}\!\!=\!\!\!\Rightarrow\!\!\underset{\alpha}{\circ}.$$

Exercise 11.7

Show that a root system is irreducible if and only if its Dynkin diagram is connected; that is, given any two vertices, there is a path joining them.

Given a Dynkin diagram, one can read off the numbers $\langle \alpha_i, \alpha_j \rangle$ and so recover the Cartan matrix. In fact, more is true: The next section shows that a root system is essentially determined by its Dynkin diagram.

11.4.1 Isomorphisms of Root Systems

Definition 11.19

Let R and R' be root systems in the real inner-product spaces E and E', respectively. We say that R and R' are isomorphic if there is a vector space isomorphism $\varphi : E \to E'$ such that

(a) $\varphi(R) = R'$, and

(b) for any two roots $\alpha, \beta \in R$, $\langle \alpha, \beta \rangle = \langle \varphi(\alpha), \varphi(\beta) \rangle$.

Recall that if θ is the angle between roots α and β, then $4\cos^2 \theta = \langle \alpha, \beta \rangle \langle \beta, \alpha \rangle$, so condition (b) says that φ should preserve angles between root vectors. For irreducible root systems, a stronger geometric characterisation is possible — see Exercise 11.15 at the end of this chapter.

Example 11.20

Let R be a root system in the inner-product space E. We used that the reflection maps s_α for $\alpha \in R$ are isomorphisms (from R to itself) when we defined the Cartan matrix of a root system.

An example of an isomorphism that is not distance preserving is given by scaling: For any non-zero $c \in \mathbf{C}$, the set $cR = \{c\alpha : \alpha \in R\}$ is a root system in E, and the map $v \mapsto cv$ induces an isomorphism between R and cR.

It follows immediately from the definition of isomorphism that isomorphic root systems have the same Dynkin diagram. We now prove that the converse holds.

Proposition 11.21

Let R and R' be root systems in the real vector spaces E and E', respectively. If the Dynkin diagrams of R and R' are the same, then the root systems are isomorphic.

Proof

We may choose bases $B = \{\alpha_1, \ldots, \alpha_\ell\}$ in R and $B' = \{\alpha_1', \ldots, \alpha_\ell'\}$ in R' so that for all i, j one has

$$\langle \alpha_i, \alpha_j \rangle = \langle \alpha_i', \alpha_j' \rangle.$$

Let $\varphi : E \to E'$ be the linear map which maps α_i to α_i'. By definition, this is a vector space isomorphism satisfying condition 11.19(b). We must show that $\varphi(R) = R'$.

Let $v \in E$ and $\alpha_i \in B$. We have

$$\varphi(s_{\alpha_i}(v)) = \varphi(v - \langle v, \alpha_i \rangle \alpha_i)$$
$$= \varphi(v) - \langle v, \alpha_i \rangle \alpha_i'.$$

We claim that $\langle v, \alpha_i \rangle = \langle \varphi(v), \alpha_i' \rangle$. To show this, express v as a linear combination of $\alpha_1, \ldots, \alpha_n$ and then use that $\langle -, - \rangle$ is linear in its first component. Therefore the last equation may be written as

$$\varphi(s_{\alpha_i}(v)) = s_{\alpha_i'}(\varphi(v)).$$

By Lemma 11.15, the simple reflections s_{α_i} generate the Weyl group of R. Hence the image under φ of the orbit of $v \in E$ under the Weyl group of R is contained in the orbit of $\varphi(v)$ under the Weyl group of R'. Now Proposition 11.14 tells us that $\{g(\alpha) : g \in W_0, \alpha \in B\} = R$ so, since $\varphi(B) = B'$, we must have $\varphi(R) \subseteq R'$.

The same argument may be applied to the inverse of φ to show that $\varphi^{-1}(R') \subseteq R$. Hence $\varphi(R) = R'$, as required. $\qquad\qquad\square$

EXERCISES

11.8.   Let E be a real inner-product space of dimension $n \geq 2$. Show that E is not the union of finitely many hyperplanes of dimension $n-1$. For a more general result, see Exercise 11.12 below.

11.9.†   Let E be a finite-dimensional real inner-product space. Let $b_1, \ldots, b_n$ be a vector space basis of E. Show that there is some $z \in E$ such that $(z, b_i) > 0$ for all i.

11.10.   Let R be as in Exercise 11.1 with $\ell = 2$. We may regard the reflections s_{α_j} for $j = 1, 2$ as linear maps on $\mathbf{R}^3$. Determine $s_{\alpha_j}(\varepsilon_i)$ for $1 \leq i \leq 3$ and $1 \leq j \leq 2$. Hence show that $W(R)$ is isomorphic to the symmetric group $\mathcal{S}_3$.

11.11.   Suppose that R is a root system in E, that R' is a root system in E', and that $\varphi : E \to E'$ is a linear map which induces an isomorphism of root systems. Show that for $\alpha \in R$ one then has

$$s_\alpha = \varphi^{-1} \circ s_{\varphi(\alpha)} \circ \varphi.$$

Prove that the Weyl group associated to R is isomorphic to the Weyl group associated to R'. (If you know what it means, prove that the pairs $(R, W(R))$ and $(R', W(R'))$ are isomorphic as G-spaces.)

11.12.†   Suppose that E is a finite-dimensional vector space over an infinite field. Suppose $U_1, U_2, \ldots, U_n$ are proper subspaces of E of the same dimension. Show that the set-theoretic union $\bigcup_{i=1}^n U_i$ is not a subspace. In particular, it is a proper subset of E.

11.13.   Suppose that R is a root system in the real inner-product space E. Show that

$$\check{R} := \left\{ \frac{2\alpha}{(\alpha, \alpha)} : \alpha \in R \right\}$$

is also a root system in E. Show that the Cartan matrix of $\check{R}$ is the transpose of the Cartan matrix of $\check{R}$ (when each is taken with respect to suitable ordering of the roots) and that the Weyl groups of R and $\check{R}$ are isomorphic. One says $\check{R}$ is the *dual root system* to R.

11.14.†   Show that if R is a root system and $\alpha, \beta \in R$ are roots with $\alpha \neq \pm\beta$ then the subgroup of the Weyl group $W(R)$ generated by s_α, s_β is a dihedral group with rotational subgroup generated by $s_\alpha s_\beta$. Hence, or otherwise, find the Weyl groups of the root systems in Example 11.6.

Hint: A group generated by two elements x and y, each of order 2, is dihedral of order $2m$, where m is the order of xy.

11.15.*   Let R and R' be irreducible root systems in the real inner-product spaces E and E'. Prove that R and R' are isomorphic if and only if there exist a scalar $\lambda \in \mathbf{R}$ and a vector space isomorphism $\varphi : E \to E'$ such that $\varphi(R) = R'$ and

$$(\varphi(u), \varphi(v)) = \lambda(u, v) \text{ for all } u, v \in E.$$

12
The Classical Lie Algebras

Our aim in this chapter is to study the classical Lie algebras $\mathsf{sl}(n, \mathbf{C})$, $\mathsf{so}(n, \mathbf{C})$, and $\mathsf{sp}(n, \mathbf{C})$ for $n \geq 2$. We shall show that, with two exceptions, all these Lie algebras are simple. We shall also find their root systems and the associated Dynkin diagrams and describe their Killing forms. The main result we prove is the following theorem.

Theorem 12.1

If L is a classical Lie algebra other than $\mathsf{so}(2, \mathbf{C})$ and $\mathsf{so}(4, \mathbf{C})$, then L is simple.

We also explain how the root systems we have determined can be used to rule out most isomorphisms between different classical Lie algebras (while suggesting the presence of those that do exist). This will lead us to a complete classification of the classical Lie algebras up to isomorphism.

In the following section, we describe a programme that will enable us to deal with each of the families of classical Lie algebras in a similar way. We then carry out this programme for each family in turn.

12.1 General Strategy

Let L be a classical Lie algebra. In each case, it follows from the definitions given in §4.3 that L has a large subalgebra H of diagonal matrices. The maps $\operatorname{ad} h$ for $h \in H$ are diagonalisable, as was first seen in Exercise 1.17, so H consists of semisimple elements.

We can immediately say a bit more about the action of H. The subspace $L \cap \operatorname{Span}\{e_{ij} : i \neq j\}$ of off-diagonal matrices in L is also invariant under $\operatorname{ad} h$ for $h \in H$ and hence the action of $\operatorname{ad} H$ on this space is diagonalisable. Let

$$L \cap \operatorname{Span}\{e_{ij} : i \neq j\} = \bigoplus_{\alpha \in \Phi} L_\alpha,$$

where for $\alpha \in H^*$, L_α is the α-eigenspace of H on the off-diagonal part of L and

$$\Phi = \{\alpha \in H^* : \alpha \neq 0, L_\alpha \neq 0\}.$$

This gives us the decomposition

$$(\star) \qquad\qquad L = L_0 \oplus \bigoplus_{\alpha \in \Phi} L_\alpha,$$

which looks very much like a root space decomposition. We shall first show that $H = L_0$, from which it will follow that H is a Cartan subalgebra of L.

Lemma 12.2

Let $L \subseteq \mathfrak{gl}(n, \mathbf{C})$ and H be as in $(\star)$ above. Suppose that for all non-zero $h \in H$ there is some $\alpha \in \Phi$ such that $\alpha(h) \neq 0$. Then H is a Cartan subalgebra of L.

Proof

We know already that H is abelian and that all the elements of H are semisimple. It remains to show that H is maximal with these properties. Suppose that $x \in L$ and that $[H, x] = 0$. (Equivalently, $x \in L_0$.)

Using the direct sum decomposition $(\star)$, we may write x as $x = h_x + \sum_{\alpha \in \Phi} c_\alpha x_\alpha$, where $x_\alpha \in L_\alpha$, $c_\alpha \in \mathbf{C}$, and $h_x \in H$. For all $h \in H$, we have

$$0 = [h, x] = \sum_\alpha c_\alpha \alpha(h) x_\alpha.$$

By the hypothesis, for every $\alpha \in \Phi$ there is some $h \in H$ such that $\alpha(h) \neq 0$, so $c_\alpha = 0$ for each α and hence $x \in H$. $\qquad\square$

To show that the classical Lie algebras (with the two exceptions mentioned in Theorem 12.1) are simple, we first need to show that they are semisimple. We shall use the following criterion.

Proposition 12.3

Let L be a complex Lie algebra with Cartan subalgebra H. Let

$$L = H \oplus \bigoplus_{\alpha \in \Phi} L_\alpha$$

be the direct sum decomposition of L into simultaneous eigenspaces for the elements of $\operatorname{ad} H$, where Φ is the set of non-zero $\alpha \in H^\star$ such that $L_\alpha \neq 0$. (So we assume that $H = L_0$.) Suppose that the following conditions hold:

(i) For each $0 \neq h \in H$, there is some $\alpha \in \Phi$ such that $\alpha(h) \neq 0$.

(ii) For each $\alpha \in \Phi$, the space L_α is 1-dimensional.

(iii) If $\alpha \in \Phi$, then $-\alpha \in \Phi$, and if L_α is spanned by x_α, then $[[x_\alpha, x_{-\alpha}], x_\alpha] \neq 0$.

Then L is semisimple.

Proof

By Exercise 4.6, it is enough to show that L has no non-zero abelian ideals. Let A be an abelian ideal of L. By hypothesis, H acts diagonalisably on L and $[H, A] \subseteq A$, so H also acts diagonalisably on A. We can therefore decompose A as

$$A = (A \cap H) \oplus \bigoplus_{\alpha \in \Phi} (A \cap L_\alpha).$$

Suppose for a contradiction that $A \cap L_\alpha \neq 0$ for some $\alpha \in \Phi$. Then, because L_α is 1-dimensional, we must have $L_\alpha \subseteq A$. Since A is an ideal, this implies that $[L_\alpha, L_{-\alpha}] \subseteq A$, so A contains an element h of the form $h = [x_\alpha, x_{-\alpha}]$, where x_α spans L_α and $x_{-\alpha}$ spans $L_{-\alpha}$. Since A is abelian and both x_α and h are known to lie in A, we deduce that $[h, x_\alpha] = 0$. However, condition (iii) says that $[h, x_\alpha] \neq 0$, a contradiction.

We have therefore proved that $A = A \cap H$; that is, $A \subseteq H$. If A contains some non-zero element h, then, by condition (i), we know that there is some $\alpha \in \Phi$ such that $\alpha(h) \neq 0$. But then $[h, x_\alpha] = \alpha(h)x_\alpha \in L_\alpha$ and also $[h, x_\alpha] \in A$, so $x_\alpha \in L_\alpha \cap A$, which contradicts the previous paragraph. Therefore $A = 0$. $\quad\square$

Note that since $[L_\alpha, L_{-\alpha}] \subseteq L_0 = H$, condition (iii) holds if and only if $\alpha([L_\alpha, L_{-\alpha}]) \neq 0$. Therefore, to show that this condition holds, it is enough to verify that $[[L_\alpha, L_{-\alpha}], L_\alpha] \neq 0$ for one member of each pair of roots $\pm\alpha$; this will help to reduce the amount of calculation required.

Having found a Cartan subalgebra of L and shown that L is semisimple, we will then attempt to identify the root system. We must find a base for Φ, and then for β, γ in the base we must find the Cartan number $\langle \beta, \gamma \rangle$. To do this, we shall use the identity

$$\langle \beta, \gamma \rangle = \beta(h_\gamma),$$

where h_γ is part of the standard basis of the subalgebra $\mathsf{sl}(\gamma)$ associated to the root γ (see §10.4). To find h_γ will be an easy calculation for which we can use the work done in checking condition (iii) of Proposition 12.3.

Now, to show that L is simple, it is enough, by the following proposition, to show that Φ is irreducible, or equivalently (by Exercise 11.7) that the Dynkin diagram of Φ is connected.

Proposition 12.4

Let L be a complex semisimple Lie algebra with Cartan subalgebra H and root system Φ. If Φ is irreducible, then L is simple.

Proof

By the root space decomposition, we may write L as

$$L = H \oplus \bigoplus_{\alpha \in \Phi} L_\alpha.$$

Suppose that L has a proper non-zero ideal I. Since H consists of semisimple elements, it acts diagonalisably on I, and so I has a basis of common eigenvectors for the elements of $\mathrm{ad}\, H$. As we know that each root space L_α is 1-dimensional, this implies that

$$I = H_1 \oplus \bigoplus_{\alpha \in \Phi_1} L_\alpha$$

for some subspace H_1 of $H = L_0$ and some subset Φ_1 of Φ. Similarly, we have

$$I^\perp = H_2 \oplus \bigoplus_{\alpha \in \Phi_2} L_\alpha,$$

where $I^\perp$ is the perpendicular space to I with respect to the Killing form. As $I \oplus I^\perp = L$, we must have $H_1 \oplus H_2 = H$, $\Phi_1 \cap \Phi_2 = \emptyset$, and $\Phi_1 \cup \Phi_2 = \Phi$.

If Φ_2 is empty, then $L_\alpha \subseteq I$ for all $\alpha \in \Phi$. As L is generated by its root spaces, this implies that $I = L$, a contradiction. Similarly, Φ_1 is non-empty. Now, given $\alpha \in \Phi_1$ and $\beta \in \Phi_2$, we have

$$\langle \alpha, \beta \rangle = \alpha(h_\beta) = 0$$

as $\alpha(h_\beta)e_\alpha = [h_\beta, e_\alpha] \in I^\perp \cap I = 0$, so $(\alpha, \beta) = 0$ for all $\alpha \in \Phi_1$ and $\beta \in \Phi_2$, which shows that Φ is reducible. $\qquad\square$

In summary, our programme is:

(1) Find the subalgebra H of diagonal matrices in L and determine the decomposition $(\star)$. This will show directly that conditions (i) and (ii) of Proposition 12.3 hold.

(2) Check that $[[L_\alpha, L_{-\alpha}], L_\alpha] \neq 0$ for each root $\alpha \in \Phi$.

By Lemma 12.2 and Proposition 12.3, we now know that L is semisimple and that H is a Cartan subalgebra of L.

(3) Find a base for Φ.

(4) For γ, β in the base, find h_γ and e_β and hence $\langle \beta, \gamma \rangle = \beta(h_\gamma)$. This will determine the Dynkin diagram of our root system, from which we can verify that Φ is irreducible and L is simple.

12.2 sl($\ell + 1, \mathbf{C}$)

For this Lie algebra, most of the work has already been done.

(1) We saw at the start of Chapter 10 that the root space decomposition of $L = \mathsf{sl}(\ell + 1, \mathbf{C})$ is

$$L = H \oplus \bigoplus_{i \neq j} L_{\varepsilon_i - \varepsilon_j},$$

where $\varepsilon_i(h)$ is the i-th entry of h and the root space $L_{\varepsilon_i - \varepsilon_j}$ is spanned by e_{ij}. Thus $\Phi = \{\pm(\varepsilon_i - \varepsilon_j) : 1 \leq i < j \leq l + 1\}$.

(2) If $i < j$, then $[e_{ij}, e_{ji}] = e_{ii} - e_{jj} = h_{ij}$ and $[h_{ij}, e_{ij}] = 2e_{ij} \neq 0$.

(3) We know from Exercise 11.4 that the root system Φ has as a base $\{\alpha_i : 1 \leq i \leq \ell\}$, where $\alpha_i = \varepsilon_i - \varepsilon_{i+1}$.

(4) From (2) we see that standard basis elements for the subalgebras $\mathsf{sl}(\alpha_i)$ can be taken as $e_{\alpha_i} = e_{i,i+1}$, $f_{\alpha_i} = e_{i+1,i}$, $h_{\alpha_i} = e_{ii} - e_{i+1,i+1}$. Calculation shows that

$$\langle \alpha_i, \alpha_j \rangle = \alpha_i(h_{\alpha_j}) = \begin{cases} 2 & i = j \\ -1 & |i - j| = 1 \\ 0 & \text{otherwise,} \end{cases}$$

so the Cartan matrix of Φ is as calculated in Example 11.17(1) and the Dynkin diagram is

This diagram is connected, so L is simple. We say that the root system of $\mathsf{sl}(\ell + 1, \mathbf{C})$ has *type* A_ℓ.

12.3 so$(2\ell + 1, \mathbf{C})$

Let $L = \mathsf{gl}_S(2\ell + 1, \mathbf{C})$ for $\ell \geq 1$, where

$$S = \begin{pmatrix} 1 & 0 & 0 \\ 0 & 0 & I_\ell \\ 0 & I_\ell & 0 \end{pmatrix}.$$

Recall that this means

$$L = \left\{ x \in \mathsf{gl}(2\ell + 1, \mathbf{C}) : x^t S = -Sx \right\}.$$

We write elements of L as block matrices, of shapes adapted to the blocks of S. Calculation shows, using Exercise 2.12, that

$$L = \left\{ \begin{pmatrix} 0 & c^t & -b^t \\ b & m & p \\ -c & q & -m^t \end{pmatrix} : p = -p^t \text{ and } q = -q^t \right\}.$$

As usual, let H be the set of diagonal matrices in L. It will be convenient to label the matrix entries from 0 to 2ℓ. Let $h \in H$ have diagonal entries $0, a_1, \ldots, a_\ell, -a_1, \ldots, -a_\ell$, so with our numbering convention,

$$h = \sum_{i=1}^{\ell} a_i(e_{ii} - e_{i+\ell,i+\ell}).$$

(1a) We start by finding the root spaces for H. Consider the subspace of L spanned by matrices whose non-zero entries occur only in the positions labelled by b and c. This subspace has as a basis $b_i = e_{i,0} - e_{0,\ell+i}$ and $c_i = e_{0,i} - e_{\ell+i,0}$ for $1 \leq i \leq \ell$. (Note that b_i and c_i are matrices, not scalars!) We calculate that

$$[h, b_i] = a_i b_i, \quad [h, c_i] = -a_i c_i.$$

(1b) We extend to a basis of L by the matrices

$$m_{ij} = e_{ij} - e_{\ell+j,\ell+i} \text{ for } 1 \leq i \neq j \leq \ell,$$
$$p_{ij} = e_{i,\ell+j} - e_{j,\ell+i} \text{ for } 1 \leq i < j \leq l,$$
$$q_{ji} = p_{ij}^t = e_{\ell+j,i} - e_{\ell+i,j} \text{ for } 1 \leq i < j \leq l.$$

Again we are fortunate that the obvious basis elements are in fact simultaneous eigenvectors for the action of H. Calculation shows that

$$[h, m_{ij}] = (a_i - a_j) m_{ij},$$
$$[h, p_{ij}] = (a_i + a_j) p_{ij},$$
$$[h, q_{ji}] = -(a_i + a_j) q_{ji}.$$

We can now list the roots. For $1 \leq i \leq \ell$, let $\varepsilon_i \in H^\star$ be the map sending h to a_i, its entry in position i.

root	ε_i	$-\varepsilon_i$	$\varepsilon_i - \varepsilon_j$	$\varepsilon_i + \varepsilon_j$	$-(\varepsilon_i + \varepsilon_j)$
eigenvector	b_i	c_i	m_{ij} $(i \neq j)$	p_{ij} $(i < j)$	q_{ji} $(i < j)$

(2) We check that $[h, x_\alpha] \neq 0$, where $h = [x_\alpha, x_{-\alpha}]$.

(2a) For $\alpha = \varepsilon_i$, we have

$$h_i := [b_i, c_i] = e_{ii} - e_{\ell+i,\ell+i}$$

and, by (1a), $[h_i, b_i] = b_i$.

(2b) For $\alpha = \varepsilon_i - \varepsilon_j$ and $i < j$, we have

$$h_{ij} := [m_{ij}, m_{ji}] = (e_{ii} - e_{\ell+i,\ell+i}) - (e_{jj} - e_{\ell+j,\ell+j})$$

and, by (1b), $[h_{ij}, m_{ij}] = 2m_{ij}$.

(2c) Finally, for $\alpha = \varepsilon_i + \varepsilon_j$, for $i < j$, we have

$$k_{ij} := [p_{ij}, q_{ji}] = (e_{ii} - e_{\ell+i,\ell+i}) + (e_{jj} - e_{\ell+j,\ell+j})$$

and, by (1b), $[k_{ij}, p_{ij}] = 2p_{ij}$.

(3) We claim that a base for our root system is given by

$$B = \{\alpha_i : 1 \leq i < \ell\} \cup \{\beta_\ell\},$$

where $\alpha_i = \varepsilon_i - \varepsilon_{i+1}$ and $\beta_\ell = \varepsilon_\ell$. To see this, note that when $1 \leq i < \ell$,

$$\varepsilon_i = \alpha_i + \alpha_{i+1} + \ldots + \alpha_{\ell-1} + \beta_\ell,$$

and that when $1 \leq i < j \leq \ell$,

$$\varepsilon_i - \varepsilon_j = \alpha_i + \alpha_{i+1} + \ldots + \alpha_{j-1},$$
$$\varepsilon_i + \varepsilon_j = \alpha_i + \ldots \alpha_{j-1} + 2(\alpha_j + \alpha_{j+1} + \ldots + \alpha_{\ell-1} + \beta_\ell).$$

Going through the table of roots shows that if $\gamma \in \Phi$ then either γ or $-\gamma$ appears above as a non-negative linear combination of elements of B. Since B has ℓ elements and $\ell = \dim H$, this is enough to show that B is a base of Φ.

(4) We now determine the Cartan matrix. For $i < \ell$, we take $e_{\alpha_i} = m_{i,i+1}$, and then $h_{\alpha_i} = h_{i,i+1}$ follows from (2b). We take $e_{\beta_\ell} = b_\ell$, and then from (2a) we see that $h_\beta = 2(e_{\ell,\ell} - e_{2\ell,2\ell})$.

For $1 \leq i, j < \ell$, we calculate that

$$[h_{\alpha_j}, e_{\alpha_i}] = \begin{cases} 2e_{\alpha_j} & i = j \\ -e_{\alpha_j} & |i - j| = 1 \\ 0 & \text{otherwise.} \end{cases}$$

Hence

$$\langle \alpha_i, \alpha_j \rangle = \begin{cases} 2 & i = j \\ -1 & |i - j| = 1 \\ 0 & \text{otherwise.} \end{cases}$$

Similarly, by calculating $[h_{\beta_\ell}, e_{\alpha_i}]$ and $[h_{\alpha_i}, e_{\beta_\ell}]$, we find that

$$\langle \alpha_i, \beta_\ell \rangle = \begin{cases} -2 & i = \ell - 1 \\ 0 & \text{otherwise,} \end{cases}$$

$$\langle \beta_\ell, \alpha_i \rangle = \begin{cases} -1 & i = \ell - 1 \\ 0 & \text{otherwise.} \end{cases}$$

This shows that the Dynkin diagram of Φ is

As the Dynkin diagram is connected, Φ is irreducible and so L is simple. The root system of $\mathbf{so}(2\ell + 1, \mathbf{C})$ is said to have *type* B_ℓ.

12.4 so(2ℓ, C)

Let $L = \mathfrak{gl}_S(2\ell, \mathbf{C})$, where

$$S = \begin{pmatrix} 0 & I_\ell \\ I_\ell & 0 \end{pmatrix}.$$

We write elements of L as block matrices, of shapes adapted to the blocks of S. Calculation shows that

$$L = \left\{ \begin{pmatrix} m & p \\ q & -m^t \end{pmatrix} : p = -p^t \text{ and } q = -q^t \right\}.$$

We see that if $\ell = 1$ then the Lie algebra is 1-dimensional, and so, by definition, not simple or semisimple. For this reason, we assumed in the statement of Theorem 12.1 that $\ell \geq 2$.

As usual, we let H be the set of diagonal matrices in L. We label the matrix entries from 1 up to 2ℓ. This means that we can use the calculations already done for $\mathfrak{so}(2\ell+1, \mathbf{C})$ by ignoring the row and column of matrices labelled by 0.

(1) All the work needed to find the root spaces in $\mathfrak{so}(2\ell, \mathbf{C})$ is done for us by step (1b) for $\mathfrak{so}(2\ell + 1, \mathbf{C})$. Taking the notation from this part, we get the following roots:

root	$\varepsilon_i - \varepsilon_j$	$\varepsilon_i + \varepsilon_j$	$-(\varepsilon_i + \varepsilon_j)$
eigenvector	$m_{ij}\ (i \neq j)$	$p_{ij}\ (i < j)$	$q_{ji}\ (i < j)$

(2) The work already done in steps (2b) and (2c) for $\mathfrak{sl}(2\ell + 1, \mathbf{C})$ shows that $[[L_\alpha, L_{-\alpha}], L_\alpha] \neq 0$ for each root α.

(3) We claim that a base for our root system is given by

$$B = \{\alpha_i : 1 \leq i < \ell\} \cup \{\beta_\ell\},$$

where $\alpha_i = \varepsilon_i - \varepsilon_{i+1}$ and $\beta_\ell = \varepsilon_{\ell-1} + \varepsilon_\ell$. To see this, note that when $1 \leq i < j < \ell$,

$$\varepsilon_i - \varepsilon_j = \alpha_i + \alpha_{i+1} + \ldots + \alpha_{j-1},$$
$$\varepsilon_i + \varepsilon_j = (\alpha_i + \alpha_{i+1} + \ldots + \alpha_{\ell-2}) + (\alpha_j + \alpha_{j+1} + \ldots + \alpha_{\ell-1} + \beta_\ell).$$

This shows that if $\gamma \in \Phi$, then either γ or $-\gamma$ is a non-negative linear combination of elements of B with integer coefficients, so B is a base for our root system.

(4) We calculate the Cartan integers. The work already done for $\mathsf{so}(2\ell+1,\mathbf{C})$ gives us the Cartan numbers $\langle \alpha_i, \alpha_j \rangle$ for $i, j < \ell$. For the remaining ones, we take $e_{\beta_\ell} = p_{\ell-1,\ell}$ and then we find from step (2c) for $\mathsf{so}(2\ell+1,\mathbf{C})$ that

$$h_{\beta_\ell} = (e_{\ell-1,\ell-1} - e_{2\ell-1,2\ell-1}) + (e_{\ell,\ell} - e_{2\ell,2\ell}).$$

Hence

$$\langle \alpha_j, \beta_\ell \rangle = \begin{cases} -1 & j = \ell - 2 \\ 0 & \text{otherwise,} \end{cases}$$

$$\langle \beta_\ell, \alpha_j \rangle = \begin{cases} -1 & j = \ell - 2 \\ 0 & \text{otherwise.} \end{cases}$$

If $\ell = 2$, then the base has only the two orthogonal roots α_1 and β_2, so in this case, Φ is reducible. In fact, $\mathsf{so}(4,\mathbf{C})$ is isomorphic to $\mathsf{sl}(2,\mathbf{C}) \oplus \mathsf{sl}(2,\mathbf{C})$, as you were asked to prove in Exercise 10.8. This explains the other Lie algebra excluded from the statement of Theorem 12.1.

If $\ell \geq 3$, then our calculation shows that the Dynkin diagram of Φ is

As this diagram is connected, the Lie algebra is simple. When $\ell = 3$, the Dynkin diagram is the same as that of A_3, the root system of $\mathsf{sl}(3,\mathbf{C})$, so we might expect that $\mathsf{so}(6,\mathbf{C})$ should be isomorphic to $\mathsf{sl}(4,\mathbf{C})$. This is indeed the case; see Exercise 14.1. For $\ell \geq 4$, the root system of $\mathsf{so}(2\ell,\mathbf{C})$ is said to have *type D_ℓ*.

12.5 sp$(2\ell, \mathbf{C})$

Let $L = \mathsf{gl}_S(2\ell, \mathbf{C})$, where S is the matrix

$$S = \begin{pmatrix} 0 & I_\ell \\ -I_\ell & 0 \end{pmatrix}.$$

We write elements of L as block matrices, of shapes adapted to the blocks of S. Calculation shows that

$$L = \left\{ \begin{pmatrix} m & p \\ q & -m^t \end{pmatrix} : p = p^t \text{ and } q = q^t \right\}.$$

We see that when $\ell = 1$, this is the same Lie algebra as sl($2, \mathbf{C}$). In what follows, we shall assume that $\ell \geq 2$.

Let H be the set of diagonal matrices in L. We label the matrix entries in the usual way from 1 to 2ℓ. Let $h \in H$ have diagonal entries $a_1, \ldots, a_\ell, -a_1, \ldots, -a_\ell$, that is,

$$h = \sum_{i=1}^{\ell} a_i (e_{ii} - e_{i+\ell, i+\ell}).$$

(1) We take the following basis for the root spaces of L:

$$m_{ij} = e_{ij} - e_{\ell+j, \ell+i} \text{ for } 1 \leq i \neq j \leq \ell,$$

$$p_{ij} = e_{i, \ell+j} + e_{j, \ell+i} \text{ for } 1 \leq i < j \leq \ell, \quad p_{ii} = e_{i, \ell+i} \text{ for } 1 \leq i \leq \ell,$$

$$q_{ji} = p_{ij}^t = e_{\ell+j, i} + e_{\ell+i, j} \text{ for } 1 \leq i < j \leq \ell, \quad q_{ii} = e_{\ell+i, i} \text{ for } 1 \leq i \leq \ell.$$

Calculation shows that

$$[h, m_{ij}] = (a_i - a_j) m_{ij},$$
$$[h, p_{ij}] = (a_i + a_j) p_{ij},$$
$$[h, q_{ji}] = -(a_i + a_j) q_{ji}.$$

Notice that for p_{ij} and q_{ji} it is allowed that $i = j$, and in these cases we get the eigenvalues $2a_i$ and $-2a_i$, respectively.

We can now list the roots. Write ε_i for the element in H^* sending h to a_i.

root	$\varepsilon_i - \varepsilon_j$	$\varepsilon_i + \varepsilon_j$	$-(\varepsilon_i + \varepsilon_j)$	$2\varepsilon_i$	$-2\varepsilon_i$
eigenvector	m_{ij} ($i \neq j$)	p_{ij} ($i < j$)	q_{ji} ($i < j$)	p_{ii}	q_{ii}

(2) For each root α, we must check that $[h, x_\alpha] \neq 0$, where $h = [x_\alpha, x_{-\alpha}]$. When $\alpha = \varepsilon_i - \varepsilon_j$, this has been done in step (2b) for so($2\ell + 1, \mathbf{C}$). If $\alpha = \varepsilon_i + \varepsilon_j$, then $x_\alpha = p_{ij}$ and $x_{-\alpha} = q_{ji}$ and

$$h = (e_{ii} - e_{\ell+i, \ell+i}) + (e_{jj} - e_{\ell+j, \ell+j})$$

if $i \neq j$, and $h = e_{ii} - e_{\ell+i, \ell+i}$ if $i = j$. Hence $[h, x_\alpha] = 2x_\alpha$ in both cases.

(3) Let $\alpha_i = \varepsilon_i - \varepsilon_{i+1}$ for $1 \leq i \leq \ell - 1$ as before, and let $\beta_\ell = 2\varepsilon_\ell$. We claim that $\{\alpha_1, \ldots, \alpha_{\ell-1}, \beta_\ell\}$ is a base for Φ. By the same argument as used before, this follows once we observe that for $1 \leq i < j \leq \ell$

$$\varepsilon_i - \varepsilon_j = \alpha_i + \alpha_{i+1} + \ldots + \alpha_{j-1},$$
$$\varepsilon_i + \varepsilon_j = \alpha_i + \alpha_{i+1} + \ldots + \alpha_{j-1} + 2(\alpha_j + \ldots + \alpha_{\ell-1}) + \beta_\ell,$$
$$2\varepsilon_i = 2(\alpha_i + \alpha_{i+1} + \ldots + \alpha_{\ell-1}) + \beta_\ell.$$

(4) We calculate the Cartan integers. The numbers $\langle \alpha_i, \alpha_j \rangle$ are already known. Take $e_{\beta_\ell} = p_{\ell\ell}$, then we find that $h_{\beta_\ell} = e_{\ell,\ell} - e_{2\ell,2\ell}$ and so

$$\langle \alpha_i, \beta_\ell \rangle = \begin{cases} -1 & i = \ell - 1 \\ 0 & \text{otherwise,} \end{cases}$$

$$\langle \beta_\ell, \alpha_j \rangle = \begin{cases} -2 & i = \ell - 1 \\ 0 & \text{otherwise.} \end{cases}$$

The Dynkin diagram of Φ is

which is connected, so L is simple. The root system of $\mathsf{sp}(2\ell, \mathbf{C})$ is said to have *type* C_ℓ. Since the root systems C_2 and B_2 have the same Dynkin diagram, we might expect that the Lie algebras $\mathsf{sp}(4, \mathbf{C})$ and $\mathsf{so}(5, \mathbf{C})$ would be isomorphic. This is the case, see Exercise 13.1.

12.6 Killing Forms of the Classical Lie Algebras

Now that we know that (with two exceptions) the classical Lie algebras are simple, we can use some of our earlier work to compute their Killing forms. We shall see that they can all be given by a simple closed formula.

Lemma 12.5

Let $L \subseteq \mathsf{gl}(n, \mathbf{C})$ be a simple classical Lie algebra. Let $\beta : L \times L \to \mathbf{C}$ be the symmetric bilinear form

$$\beta(x, y) := \text{tr}(xy).$$

Then β is non-degenerate.

Proof

Let $J = \{x \in L : \beta(x, y) = 0 \text{ for all } y \in L\}$. It follows from the associative property of trace, as in Exercise 9.3, that J is an ideal of L. Since L is simple, and clearly β is not identically zero, we must have $J = 0$. Therefore β is non-degenerate. $\qquad \square$

In Exercise 9.11, we showed that any two non-degenerate symmetric associative bilinear forms on a simple Lie algebra are scalar multiples of one another.

Hence, by Cartan's Second Criterion (Theorem 9.9), if κ is the Killing form on L, then $\kappa = \lambda\beta$ for some non-zero scalar $\lambda \in \mathbf{C}$. To determine the scalar λ, we use the root space decomposition to compute $\kappa(h, h')$ for $h, h' \in H$. For example, for $\mathsf{sl}(\ell+1, \mathbf{C})$ let $h \in H$, with diagonal entries $a_1, \ldots, a_{\ell+1}$, and similarly let $h' \in H$ with diagonal entries $a'_1, \ldots, a'_{\ell+1}$. Then, using the root space decomposition given in step (1) of §12.2, we get

$$\kappa(h, h') = \sum_{\alpha \in \Phi} \alpha(h)\alpha(h') = 2\sum_{i<j}(a_i - a_j)(a'_i - a'_j).$$

Putting $h = h'$, and $a_1 = 1$, $a_2 = -1$ and all other entries zero, we get $\kappa(h, h) = 8 + 4(\ell - 1) = 4(\ell + 1)$. Since $\operatorname{tr} h^2 = 2$, this implies that $\lambda = 2(\ell+1)$.

For the remaining three families, see Exercise 12.3 below (or its solution in Appendix E). We get $\kappa(x, y) = \lambda \operatorname{tr}(xy)$, where

$$\lambda = \begin{cases} 2(\ell+1) & L = \mathsf{sl}(\ell+1, \mathbf{C}) \\ 2\ell - 1 & L = \mathsf{so}(2\ell+1, \mathbf{C}) \\ 2(\ell+1) & L = \mathsf{sp}(2\ell, \mathbf{C}) \\ 2(\ell-1) & L = \mathsf{so}(2\ell, \mathbf{C}). \end{cases}$$

12.7 Root Systems and Isomorphisms

Let L be a complex semisimple Lie algebra. We have seen how to define the root system associated to a Cartan subalgebra of L. Could two different Cartan subalgebras of L give different root systems? The following theorem, whose proof may be found in Appendix C, shows that the answer is no.

Theorem 12.6

Let L be a complex semisimple Lie algebra. If Φ_1 and Φ_2 are the root systems associated to two Cartan subalgebras of L, then Φ_1 is isomorphic to Φ_2. $\qquad\square$

Suppose now that L_1 and L_2 are complex semisimple Lie algebras that have non-isomorphic root systems (with respect to some Cartan subalgebras). Then, by the theorem, L_1 and L_2 cannot be isomorphic. Thus we can use root systems to rule out isomorphisms between the classical Lie algebras. This does most of the work needed to prove the following proposition.

Proposition 12.7

The only isomorphisms between classical Lie algebras are:

(1) $so(3, \mathbf{C}) \cong sp(2, \mathbf{C}) \cong sl(2, \mathbf{C})$; root systems of type A_1,

(2) $so(4, \mathbf{C}) \cong sl(2, \mathbf{C}) \oplus sl(2, \mathbf{C})$; root systems of type $A_1 \times A_1$,

(3) $so(5, \mathbf{C}) \cong sp(4, \mathbf{C})$; root systems of types B_2 and C_2,

(4) $so(6, \mathbf{C}) \cong sl(4, \mathbf{C})$; root systems of types D_3 and A_3.

Note that we have not yet proved the existence of all these isomorphisms. However, we have already seen the first two (see Exercises 1.14 and 10.8). The third isomorphism appears in Exercise 12.2 below and the last is discussed in Chapter 15. We are therefore led to conjecture that the converse of Theorem 12.6 also holds; that is, if two complex semisimple Lie algebras have isomorphic root systems, then they are isomorphic as Lie algebras.

We shall see in Chapter 14 that this is a corollary of Serre's Theorem. Thus isomorphisms of root systems precisely reflect isomorphisms of complex semisimple Lie algebras. To classify the complex semisimple Lie algebras, we should therefore first classify root systems. This is the subject of the next chapter.

EXERCISES

12.1.   Show that the dimensions of the classical Lie algebras are as follows

$$\dim sl(\ell + 1, \mathbf{C}) = \ell^2 + 2\ell,$$
$$\dim so(2\ell + 1, \mathbf{C}) = 2\ell^2 + \ell,$$
$$\dim sp(2\ell, \mathbf{C}) = 2\ell^2 + \ell,$$
$$\dim so(2\ell, \mathbf{C}) = 2\ell^2 - \ell.$$

12.2.* Show that the Lie algebras $sp(4, \mathbf{C})$ and $so(5, \mathbf{C})$ are isomorphic. (For instance, use the root space decomposition to show that they have bases affording the same structure constants.)

12.3.† This exercise gives a way to establish the semisimplicity of the classical Lie algebras using the Killing form.

   (i) Let L be a classical Lie algebra, and let H be the subalgebra of diagonal matrices, with eigenspace decomposition

$$L = H \oplus \bigoplus_{\alpha \in \Phi} L_\alpha,$$

so H is self-centralising. Assume also that the following conditions hold

(a) For each $\alpha \in \Phi$, the space L_α is 1-dimensional. If $\alpha \in \Phi$, then $-\alpha \in \Phi$.

(b) For each $\alpha \in \Phi$, the space $[L_\alpha, L_{-\alpha}]$ is non-zero.

(c) The Killing form restricted to H is non-degenerate, and for $h \in H$, if $\kappa(h, h) = 0$ then $h = 0$.

Show that the Killing form of L is then non-degenerate.

In the earlier sections of this chapter, we have found the roots with respect to H explicitly. We can make use of this and find the Killing form restricted to H explicitly.

(ii) Use the root space decomposition of $\mathsf{sl}(\ell + 1, \mathbf{C})$ to show that if κ is the Killing form of $\mathsf{sl}(\ell + 1, \mathbf{C})$, then

$$\kappa(h, h') = 2n \operatorname{tr}(hh') \quad \text{for all } h, h' \in H.$$

Hence, show that condition (c) above holds for the restriction of κ to H and deduce that $\mathsf{sl}(\ell + 1, \mathbf{C})$ is semisimple.

(iii) Use similar methods to prove that the orthogonal and symplectic Lie algebras are semisimple.

12.4.†* Let L be a Lie algebra with a faithful irreducible representation. Show that either L is semisimple or $Z(L)$ is 1-dimensional and $L = Z(L) \oplus L'$, where the derived algebra L' is semisimple. (This gives yet another way to prove the semisimplicity of the classical Lie algebras.)

13
The Classification of Root Systems

In §11.4, we saw how to define the Dynkin diagram of a root system. By Theorem 11.16, which states that any two bases of a root system are conjugate by an element of the Weyl group, this diagram is unique up to the labelling of the vertices. (The labels merely indicate our notation for elements of the base; and so have no essential importance.) Conversely, we saw in §11.4.1 that a root system is determined up to isomorphism by its Dynkin diagram.

From the point of view of classifying complex semisimple Lie algebras, there is no need to distinguish between isomorphic root systems. Hence the problem of finding all root systems can be reduced to the problem of finding all Dynkin diagrams; this gives us a very convenient way to organise the classification.

We shall prove that apart from the four infinite families of root systems associated to the classical Lie algebras there are just five more root systems, the so-called exceptional root systems. We end this chapter by saying a little about how they may be constructed.

13.1 Classification of Dynkin Diagrams

Our aim in this section is to prove the following theorem.

Theorem 13.1

Given an irreducible root system R, the unlabelled Dynkin diagram associated to R is either a member of one of the four families

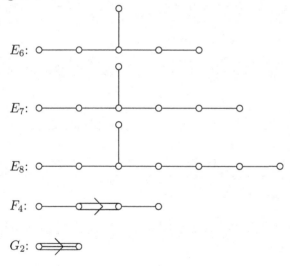

where each of the diagrams above has ℓ vertices, or one of the five exceptional diagrams

E_6:

E_7:

E_8:

F_4:

G_2:

Note that there are no repetitions in this list. For example, we have not included C_2 in the list, as it is the same diagram as B_2, and so the associated root systems are isomorphic. (Exercise 13.1 at the end of this chapter asks you to construct an explicit isomorphism.)

Let Δ be a connected Dynkin diagram. As a first approximation, we shall determine the possible underlying graphs for Δ, ignoring for the moment any arrows that may appear. To find these graphs, we do not need to know that they come from root systems. Instead it is convenient to work with more general sets of vectors.

Definition 13.2

Let E be a real inner-product space with inner product $(-, -)$. A subset A of E consisting of linearly independent vectors $v_1, v_2, \ldots, v_n$ is said to be *admissible* if it satisfies the following conditions:

(a) $(v_i, v_i) = 1$ for all i and $(v_i, v_j) \leq 0$ if $i \neq j$.

(b) If $i \neq j$, then $4(v_i, v_j)^2 \in \{0, 1, 2, 3\}$.

To the admissible set A, we associate the graph Γ_A with vertices labelled by the vectors $v_1, \ldots, v_n$, and with $d_{ij} := 4(v_i, v_j)^2 \in \{0, 1, 2, 3\}$ edges between v_i and v_j for $i \neq j$.

Example 13.3

Suppose that B is a base of a root system. Set $A := \{\alpha/\sqrt{(\alpha, \alpha)} : \alpha \in B\}$. Then A is easily seen to be an admissible set. Moreover, the graph Γ_A is the Coxeter graph of B, as defined in §11.4.

We now find all the connected graphs that correspond to admissible sets. Let A be an admissible set in the real inner-product space E with connected graph $\Gamma = \Gamma_A$. The first easy observation we make is that any subset of A is also admissible. We shall use this several times below.

Lemma 13.4

The number of pairs of vertices joined by at least one edge is at most $|A| - 1$.

Proof

Suppose $A = \{v_1, \ldots, v_n\}$. Set $v = \sum_{i=1}^n v_i$. As A is linearly independent, $v \neq 0$. Hence $(v, v) = n + 2\sum_{i<j}(v_i, v_j) > 0$ and so

$$n > \sum_{i<j} -2(v_i, v_j) = \sum_{i<j} \sqrt{d_{ij}} \geq N,$$

where N is the number of pairs $\{v_i, v_j\}$ such that $d_{ij} \geq 1$; this is the number that interests us. $\qquad\square$

Corollary 13.5

The graph Γ does not contain any cycles.

Proof

Suppose that Γ does have a cycle. Let A' be the subset of A consisting of the vectors involved in this cycle. Then A' is an admissible set with the same number (or more) edges as vertices, in contradiction to the previous lemma. □

Lemma 13.6

No vertex of Γ is incident to four or more edges.

Proof

Take a vertex v of Γ, and let $v_1, v_2, \ldots, v_k$ be all the vertices in Γ joined to v. Since Γ does not contain any cycles, we must have $(v_i, v_j) = 0$ for $i \neq j$. Consider the subspace U with basis $v_1, v_2, \ldots, v_k, v$. The Gram–Schmidt process allows us to extend $v_1, \ldots, v_k$ to an orthonormal basis of U, say by adjoining v_0; necessarily $(v, v_0) \neq 0$. We may express v in terms of this orthonormal basis as

$$v = \sum_{i=0}^{k} (v, v_i) v_i.$$

By assumption, v is a unit vector, so expanding (v, v) gives $1 = (v, v) = \sum_{i=0}^{k} (v, v_i)^2$. Since $(v, v_0)^2 > 0$, this shows that

$$\sum_{i=1}^{k} (v, v_i)^2 < 1.$$

Now, as A is admissible and $(v, v_i) \neq 0$, we know that $(v, v_i)^2 \geq \frac{1}{4}$ for $1 \leq i \leq k$. Hence $k \leq 3$.   □

An immediate corollary of this lemma is the following.

Corollary 13.7

If Γ is connected and has a triple edge, then $\Gamma = \ \text{o}\!=\!=\!=\!\text{o}$.   □

Lemma 13.8 (Shrinking Lemma)

Suppose Γ has a subgraph which is a *line*, that is, of the form

$$
\underset{v_1}{\circ}\!\!-\!\!-\!\!\underset{v_2}{\circ}\!\!-\!\!-\!\!-\quad\cdots\quad-\!\!-\!\!-\!\!\underset{v_k}{\circ}
$$

where there are no multiple edges between the vertices shown. Define $A' = (A \setminus \{v_1, v_2, \ldots, v_k\}) \cup \{v\}$ where $v = \sum_{i=1}^{k} v_i$. Then A' is admissible and the graph $\Gamma_{A'}$ is obtained from Γ_A by shrinking the line to a single vertex.

Proof

Clearly A' is linearly independent, so we need only verify the conditions on the inner products. By assumption, we have $2(v_i, v_{i+1}) = -1$ for $1 \le i \le k-1$ and $(v_i, v_j) = 0$ for $i \neq j$ otherwise. This allows us to calculate (v, v). We find that

$$
(v, v) = k + 2 \sum_{i=1}^{k-1} (v_i, v_{i+1}) = k - (k-1) = 1.
$$

Suppose that $w \in A$ and $w \neq v_i$ for $1 \le i \le k$. Then w is joined to at most one of $v_1, \ldots, v_k$ (otherwise there would be a cycle). Therefore either $(w, v) = 0$ or $(w, v) = (w, v_i)$ for some $1 \le i \le k$ and then $4(w, v)^2 \in \{0, 1, 2, 3\}$, so A' satisfies the defining conditions for an admissible set. These remarks also determine the graph $\Gamma_{A'}$. $\qquad\square$

Say that a vertex of Γ is a *branch vertex* if it is incident to three or more edges; by Lemma 13.6 such a vertex is incident to exactly three edges.

Lemma 13.9

The graph Γ has

(i) no more than one double edge;

(ii) no more than one branch vertex; and

(iii) not both a double edge and a branch vertex.

Proof

Suppose Γ has two (or more) double edges. Since Γ is connected, it has a subgraph consisting of two double edges connected by a line of the form

By the Shrinking Lemma, we obtain an admissible set with graph

which contradicts Lemma 13.6. The proofs of the remaining two parts are very similar, so we leave them to the reader. □

For the final steps of the proof of Theorem 13.1, we shall need the following calculation of an inner product.

Lemma 13.10

Suppose that Γ has a line as a subgraph:

$$v_1 \quad v_2 \qquad\qquad\qquad v_p$$

Let $v = \sum_{i=1}^{p} iv_i$. Then $(v, v) = \frac{p(p+1)}{2}$.

Proof

The shape of the subgraph tells us that $2(v_i, v_{i+1}) = -1$ for $1 \le i \le p-1$ and that $(v_i, v_j) = 0$ for $i \ne j$ otherwise, so

$$(v, v) = \sum_{i=1}^{p} i^2 + 2\sum_{i=1}^{p-1}(v_i, v_{i+1})i(i+1) = \sum_{i=1}^{p} i^2 - \sum_{i=1}^{p-1} i(i+1) = p^2 - \sum_{i=1}^{p-1} i,$$

which is equal to $p(p+1)/2$. □

Proposition 13.11

If Γ has a double edge, then Γ is one of

Proof

By Lemma 13.9, any such Γ has the form

$$v_1 \qquad\qquad\qquad v_p \quad w_q \qquad\qquad\qquad w_1$$

where, without loss of generality, $p \ge q$. Let $v = \sum_{i=1}^{p} iv_i$ and $w = \sum_{i=1}^{q} iw_i$. By the calculation above, we have

$$(v, v) = \frac{p(p+1)}{2}, \quad (w, w) = \frac{q(q+1)}{2}.$$

We see from the graph that $4(v_p, w_q)^2 = 2$ and $(v_i, w_j) = 0$ in all other cases. Hence

$$(v, w)^2 = (pv_p, qw_q)^2 = \frac{p^2 q^2}{2}.$$

As v and w are linearly independent, the Cauchy–Schwarz inequality implies that $(v, w)^2 < (v, v)(w, w)$. Substituting, we get $2pq < (p+1)(q+1)$, and hence

$$(p - 1)(q - 1) = pq - p - q + 1 < 2$$

so either $q = 1$ or $p = q = 2$.   □

Proposition 13.12

If Γ has a branch point, then either Γ is D_n for some $n \geq 4$ or Γ is E_6, E_7, or E_8.

Proof

By Lemma 13.9, any such Γ has the form

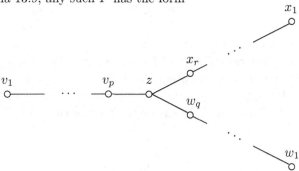

where, without loss of generality, $p \geq q \geq r$. We must show that either $q = r = 1$ or $q = 2$, $r = 1$, and $p \leq 4$.

As in the proof of the last proposition, we let $v = \sum_{i=1}^{p} iv_i$, $w = \sum_{i=1}^{q} iw_i$, and $x = \sum_{i=1}^{r} ix_i$. Then v, w, x are pairwise orthogonal. Let $\hat{v} = v/\|v\|$, $\hat{w} = w/\|w\|$, and $\hat{x} = x/\|x\|$. The space U spanned by v, w, x, z has as an orthonormal basis

$$\{\hat{v}, \hat{w}, \hat{x}, z_0\}$$

for some choice of z_0 which will satisfy $(z, z_0) \neq 0$. We may write

$$z = (z, \hat{v})\hat{v} + (z, \hat{w})\hat{w} + (z, \hat{x})\hat{x} + (z, z_0)z_0.$$

As z is a unit vector and $(z, z_0) \neq 0$, we get

$$(z, \tilde{v})^2 + (z, \tilde{w})^2 + (z, \tilde{x})^2 < 1.$$

We know the lengths of v, w, x from Lemma 13.10. Furthermore, $(z, v)^2 = (z, pv_p)^2 = p^2/4$, and similarly $(z, w)^2 = q^2/4$ and $(z, x)^2 = r^2/4$. Substituting these into the previous inequality gives

$$\frac{2p^2}{4p(p+1)} + \frac{2q^2}{4q(q+1)} + \frac{2r^2}{4r(r+1)} < 1.$$

By elementary steps, this is equivalent to

$$\frac{1}{p+1} + \frac{1}{q+1} + \frac{1}{r+1} > 1.$$

Since $\frac{1}{p+1} \leq \frac{1}{q+1} \leq \frac{1}{r+1} \leq \frac{1}{2}$, we have $1 < \frac{3}{r+1}$ and hence $r < 2$, so we must have $r = 1$. Repeating this argument gives that $q < 3$, so $q = 1$ or $q = 2$. If $q = 2$, then we see that $p < 5$. On the other hand, if $q = 1$, then there is no restriction on p. $\square$

We have now found all connected graphs which come from admissible sets. We return to our connected Dynkin diagram Δ. We saw in Example 13.3 that the Coxeter graph of Δ, say $\bar{\Delta}$, must appear somewhere in our collection. If Δ has no multiple edges, then, by Proposition 13.12, $\Delta = \bar{\Delta}$ is one of the graphs listed in Theorem 13.1.

If Δ has a double edge, then Proposition 13.11 tells us that there are two possibilities for $\bar{\Delta}$. In the case of B_2 and F_4, we get essentially the same graph whichever way we put the arrow; otherwise there are two different choices, giving B_n and C_n for $n \geq 3$. Finally, if Δ has a triple edge, then Corollary 13.7 tells us that $\Delta = G_2$. This completes the proof of Theorem 13.1.

13.2 Constructions

We now want to show that all the Dynkin diagrams listed in Theorem 13.1 actually occur as the Dynkin diagram of some root system.

Our analysis of the classical Lie algebras $\mathsf{sl}_{\ell+1}$, $\mathsf{so}_{2\ell+1}$, $\mathsf{sp}_{2\ell}$, and $\mathsf{so}_{2\ell}$ in Chapter 12 gives us constructions of root systems of types A, B, C, and D respectively. We discuss the Weyl groups of these root systems in Appendix D. For the exceptional Dynkin diagrams G_2, F_4, E_6, E_7, and E_8, we have to do more work. For completeness, we give constructions of all the corresponding root systems, but as those of type E are rather large and difficult to work with, we do not go into any details for this type.

In each case, we shall take for the underlying space E a subspace of a Euclidean space $\mathbf{R}^m$. Let ε_i be the vector with 1 in position i and 0 elsewhere.

When describing bases, we shall follow the pattern established in Chapter 12 by taking as many simple roots as possible from the set $\{\alpha_1, \ldots, \alpha_{m-1}\}$, where

$$\alpha_i := \varepsilon_i - \varepsilon_{i+1}.$$

For these elements, we have

$$\langle \alpha_i, \alpha_j \rangle = \begin{cases} 2 & i = j \\ -1 & |i - j| = 1 \\ 0 & \text{otherwise,} \end{cases}$$

so the corresponding part of the Dynkin diagram is a line,

$$\cdots \overset{\alpha_{i-1}}{\underset{}{-\!\!\!-}} \overset{\alpha_i}{\circ\!\!-\!\!-} \overset{\alpha_{i+1}}{\circ\!\!-\!\!-\circ} \cdots$$

and the corresponding part of the Cartan matrix is

$$\begin{pmatrix} & \vdots & \vdots & \vdots & \\ \cdots & 2 & -1 & 0 & \cdots \\ \cdots & -1 & 2 & -1 & \cdots \\ \cdots & 0 & -1 & 2 & \cdots \\ & \vdots & \vdots & \vdots & \end{pmatrix}.$$

Both of these will be familiar from root systems of type A.

13.2.1 Type G_2

We have already given one construction of a root system of type G_2 in Example 11.6(c). We give another here, which is more typical of the other constructions that follow. Let $E = \left\{ v = \sum_{i=1}^{3} c_i \varepsilon_i \in \mathbf{R}^3 : \sum c_i = 0 \right\}$, let

$$I = \left\{ m_1 \varepsilon_1 + m_2 \varepsilon_2 + m_3 \varepsilon_3 \in \mathbf{R}^3 : m_1, m_2, m_3 \in \mathbf{Z} \right\},$$

and let

$$R = \left\{ \alpha \in I \cap E : (\alpha, \alpha) = 2 \text{ or } (\alpha, \alpha) = 6 \right\}.$$

This is motivated by noting that the ratio of the length of a long root to the length of a short root in a root system of type G_2 is $\sqrt{3}$. By direct calculation, one finds that

$$R = \left\{ \pm(\varepsilon_i - \varepsilon_j), i \neq j \right\} \cup \left\{ \pm(2\varepsilon_i - \varepsilon_j - \varepsilon_k), \{i, j, k\} = \{1, 2, 3\} \right\}.$$

This gives 12 roots in total, as expected from the diagram in Example 11.6(c). To find a base, we need to find $\alpha, \beta \in R$ of different lengths, making an angle of $5\pi/6$. One suitable choice is $\alpha = \varepsilon_1 - \varepsilon_2$ and $\beta = \varepsilon_2 + \varepsilon_3 - 2\varepsilon_1$.

The Weyl group for G_2 is generated by the simple reflections s_α and s_β. By Exercise 11.14(ii), it is the dihedral group of order 12.

13.2.2 Type F_4

Since the Dynkin diagram F_4 contains the Dynkin diagram B_3, we might hope to construct the corresponding root system by extending the root system of B_3. Therefore we look for $\beta \in \mathbf{R}^4$ so that $B = (\varepsilon_1 - \varepsilon_2, \varepsilon_2 - \varepsilon_3, \varepsilon_3, \beta)$ is a base with Cartan numbers given by the labelled Dynkin diagram:

$$\underset{\varepsilon_1-\varepsilon_2}{\circ} \underset{\varepsilon_2-\varepsilon_3}{\circ}\!\!\!\Rightarrow\!\!\! \underset{\varepsilon_3}{\circ} \underset{\beta}{\circ}$$

It is easy to see that the only possible choices for β are $\beta = -\frac{1}{2}(\varepsilon_1 + \varepsilon_2 + \varepsilon_3) \pm \frac{1}{2}\varepsilon_4$. Therefore it seems hopeful to set

$$R = \{\pm\varepsilon_i : 1 \le i \le 4\} \cup \{\pm\varepsilon_i \pm \varepsilon_j : 1 \le i \ne j \le 4\} \cup \{\tfrac{1}{2}(\pm\varepsilon_1 \pm \varepsilon_2 \pm \varepsilon_3 \pm \varepsilon_4)\}.$$

One can check directly that axioms (R1) up to (R4) hold; see Exercise 13.3. It remains to check that

$$\beta_1 = \varepsilon_1 - \varepsilon_2,$$
$$\beta_2 = \varepsilon_2 - \varepsilon_3,$$
$$\beta_3 = \varepsilon_3,$$
$$\beta_4 = \frac{1}{2}(-\varepsilon_1 - \varepsilon_2 - \varepsilon_3 + \varepsilon_4),$$

defines a base for R. Note that R has 48 elements, so we need to find 24 positive roots. Each ε_i is a positive root, and if $1 \le i < j \le 3$ then so are $\varepsilon_i - \varepsilon_j$ and $\varepsilon_i + \varepsilon_j$. Furthermore, for $1 \le i \le 3$, also $\varepsilon_4 \pm \varepsilon_i$ are positive roots.

This already gives us 16 roots. In total, there are 16 roots of the form $\frac{1}{2}(\sum \pm\varepsilon_i)$. As one would expect, half of these turn out to be positive roots. Obviously each must have a summand equal to β_4. There are 3 positive roots of the form $\beta_4 + \varepsilon_j$, and also 3 of the form $\beta_4 + \varepsilon_j + \varepsilon_k$. Then there is β_4 itself, and finally $\beta_4 + \varepsilon_1 + \varepsilon_2 + \varepsilon_3 = \frac{1}{2}\sum \varepsilon_i$.

The Weyl group is known to have order $2^7\, 3^2$, but its structure is too complicated to be discussed here.

13.2.3 Type E

To construct the root systems of types E, it will be convenient to first construct a root system of type E_8 and then to find root systems of types E_6 and E_7 inside it.

Let $E = \mathbf{R}^8$ and let

$$R = \left\{\pm\varepsilon_i \pm \varepsilon_j : i < j\right\} \cup \left\{\frac{1}{2}\sum_{i=1}^{8} \pm\varepsilon_i\right\},$$

where in the second set an even number of + signs are chosen.

The first set in the union contributes 112 roots and the second 128, giving 240 roots in all. Assuming that R is a root system, we claim that a base for R is given by $B = \{\beta_1, \beta_2, \ldots, \beta_8\}$, where

$$\beta_1 = \frac{1}{2}\left(-\varepsilon_1 - \varepsilon_8 + \sum_{i=2}^{7} \varepsilon_i\right),$$

$$\beta_2 = -\varepsilon_1 - \varepsilon_2,$$

$$\beta_i = \varepsilon_{i-2} - \varepsilon_{i-1} \text{ if } 3 \leq i \leq 8.$$

To see that B is a base, one first verifies that the roots $\pm\varepsilon_i - \varepsilon_j$ for $i < j$ can be written as linear combinations of the elements of B with positive coefficients. The remaining positive roots are those of the form

$$\frac{1}{2}\left(-\varepsilon_8 + \sum_{i=1}^{7} \pm\varepsilon_i\right),$$

where there are an odd number of + signs chosen in the sum. To check this, subtract off β_1, and then verify that the result is a positive integral linear combination of the remaining roots. (This is where the condition on the signs comes in.) The labelled Dynkin diagram is

Omitting the root β_8 gives a base for a root system of type E_7, and omitting the roots β_7 and β_8 gives a base for a root system of type E_6. We leave it to the keen reader to explicitly construct these root systems. (Details may be found in Bourbaki, *Lie Groups and Lie Algebras* [6], Chapter 5, Section 4, Number 4.)

EXERCISES

13.1. Find an explicit isomorphism between the root systems of types B_2 and C_2. That is, find a linear map between the vector spaces for B_2 and C_2, respectively, which interchanges a long root with a short root, and preserves the Cartan numbers.

13.2. Show that the root systems of types B_n and C_n are dual to one another in the sense defined in Exercise 11.13.

13.3.   Check that the construction of F_4 given in §13.2.2 really does give a
root system. This can be simplified by noting that R contains

$$\{\pm\varepsilon_i : 1 \leq i \leq 3\} \cup \{\pm\varepsilon_i \pm \varepsilon_j : 1 \leq i \neq j \leq 3\},$$

which is the root system of type B_3 we constructed in Chapter 12.

Simple Lie Algebras

In this chapter, we shall show that for each isomorphism class of irreducible root systems there is a unique simple Lie algebra over $\mathbf{C}$ (up to isomorphism) with that root system. Moreover, we shall prove that every simple Lie algebra has an irreducible root system, so every simple Lie algebra arises in this way. These results mean that the classification of irreducible root systems in Chapter 13 gives us a complete classification of all complex simple Lie algebras.

We have already shown in Proposition 12.4 that if the root system of a Lie algebra is irreducible, then the Lie algebra is simple. We now show that the converse holds; that is, the root system of a simple Lie algebra is irreducible. We need the following lemma concerning reducible root systems.

Lemma 14.1

Suppose that Φ is a root system and that $\Phi = \Phi_1 \cup \Phi_2$ where $(\alpha, \beta) = 0$ for all $\alpha \in \Phi_1$, $\beta \in \Phi_2$.

(a) If $\alpha \in \Phi_1$ and $\beta \in \Phi_2$, then $\alpha + \beta \notin \Phi$.

(b) If $\alpha, \alpha' \in \Phi_1$ and $\alpha + \alpha' \in \Phi$, then $\alpha + \alpha' \in \Phi_1$.

Proof

For (a), note that $(\alpha, \alpha + \beta) = (\alpha, \alpha) \neq 0$, so $\alpha + \beta \notin \Phi_2$. Similarly, $(\beta, \alpha + \beta) = (\beta, \beta) \neq 0$, so $\alpha + \beta \notin \Phi_1$.

To prove (b), we suppose for a contradiction that $\alpha + \alpha' \in \Phi_2$. Remembering that $-\alpha' \in \Phi_1$, we have $\alpha = -\alpha' + (\alpha + \alpha')$, so α can be expressed as the sum of a root in Φ_1 and a root in Φ_2. This contradicts the previous part. $\square$

Proposition 14.2

Let L be a complex semisimple Lie algebra with Cartan subalgebra H and root system Φ. If L is simple, then Φ is irreducible.

Proof

By the root space decomposition, we may write L as

$$L = H \oplus \bigoplus_{\alpha \in \Phi} L_\alpha.$$

Suppose that Φ is reducible, with $\Phi = \Phi_1 \cup \Phi_2$, where Φ_1 and Φ_2 are non-empty and $(\alpha, \beta) = 0$ for all $\alpha \in \Phi_1$ and $\beta \in \Phi_2$. We shall show that the root spaces L_α for $\alpha \in \Phi_1$ generate a proper ideal of L, and so L is not simple.

For each $\alpha \in \Phi_1$ we have defined a Lie subalgebra $\mathsf{sl}(\alpha) \cong \mathsf{sl}(2, \mathbf{C})$ of L with standard basis $\{e_\alpha, f_\alpha, h_\alpha\}$. Let

$$I := \mathrm{Span}\{e_\alpha, f_\alpha, h_\alpha : \alpha \in \Phi_1\}.$$

The root space decomposition shows that I is a non-zero proper subspace of L.

We claim that I is an ideal of L; it is a subspace by definition, so we only have to show that $[x, a] \in L$ for all $x \in L$ and $a \in I$. For this it suffices to take $a = e_\alpha$ and $a = f_\alpha$ for $\alpha \in \Phi_1$ since these elements generate I. Moreover, we may assume that x lies in one of the summands of the root space decomposition of L.

If $x \in H$, then $[x, e_\alpha] = \alpha(x)e_\alpha \in I$ and similarly $[x, f_\alpha] = -\alpha(x)e_\alpha \in I$. Suppose that $x \in L_\beta$. Then, for any $\alpha \in \Phi_1$, $[x, e_\alpha] \in L_{\alpha+\beta}$ by Lemma 10.1(i). If $\beta \in \Phi_2$, then by Lemma 14.1(a) above, we know that $\alpha + \beta$ is not a root, so $L_{\alpha+\beta} = 0$, and hence $[x, e_\alpha] \in I$. Otherwise $\beta \in \Phi_1$, and then by Lemma 14.1(b) we know that $\alpha + \beta \in \Phi_1$, so $L_{\alpha+\beta} \subseteq I$, by the definition of I. Similarly, one shows that $[x, f_\alpha] \in I$. (Alternatively, one may argue that as f_α is a scalar multiple of $e_{-\alpha}$, it is enough to look at the elements e_α.) $\square$

14.1 Serre's Theorem

Serre's Theorem is a way to describe a complex semisimple Lie algebra by generators and relations that depend only on data from its Cartan matrix. The reader will probably have seen examples of groups, such as the dihedral groups, given by specifying a set of generators and the relations that they satisfy. The situation for Lie algebras is analogous.

14.1.1 Generators

Let L be a complex semisimple Lie algebra with Cartan subalgebra H and root system Φ. Suppose that Φ has as a base $\{\alpha_1, \ldots, \alpha_\ell\}$. For each i between 1 and ℓ let e_i, f_i, h_i be a standard basis of $\mathsf{sl}(\alpha_i)$. We ask whether the e_i, f_i, h_i for $1 \leq i \leq \ell$ might already generate L; that is, can every element of L be obtained by repeatedly taking linear combinations and Lie brackets of these elements?

Example 14.3

Let $L = \mathsf{sl}(\ell + 1, \mathbf{C})$. We shall show that the elements $e_{i,i+1}$ and $e_{i+1,i}$ for $1 \leq i \leq \ell$ already generate L as a Lie algebra. By taking the commutators $[e_{i,i+1}, e_{i+1,i}]$ we get a basis for the Cartan subalgebra H of diagonal matrices. For $i + 1 < j$, we have $[e_{i,i+1}, e_{i+1,j}] = e_{ij}$, and hence by induction we get all e_{ij} with $i < j$. Similarly, we may obtain all e_{ij} with $i > j$.

It is useful to look at these in terms of roots. Recall that the root system of L with respect to H has as a base $\alpha_1, \ldots, \alpha_\ell$, where $\alpha_i = \varepsilon_i - \varepsilon_{i+1}$. For $i < j$, we have $\mathrm{Span}\{e_{ij}\} = L_\beta$, where $\beta = \alpha_i + \gamma$ and $\gamma = \alpha_{i+1} + \ldots + \alpha_{j-1}$. This can be expressed neatly using reflections since

$$s_{\alpha_i}(\gamma) = \gamma - \langle \gamma, \alpha_i \rangle \, \alpha_i = \gamma + \alpha_i = \beta.$$

In fact, this method gives a general way to obtain any non-zero root space. To show this, we only need to remind ourselves of some earlier results.

Lemma 14.4

Let L be a complex semisimple Lie algebra, and let $\{\alpha_1, \ldots, \alpha_\ell\}$ be a base of the root system. Suppose $\{e_i, f_i, h_i\}$ is a standard basis of $\mathsf{sl}(\alpha_i)$. Then L can be generated, as a Lie algebra, by $\{e_1, \ldots, e_\ell, f_1, \ldots, f_\ell\}$.

Proof

We first show that every element of H can be obtained. Since $h_i = [e_i, f_i]$, it is sufficient to prove that H is spanned by $h_1, \ldots, h_\ell$. Recall that we have identified H with H^*, via the Killing form κ, so that $\alpha_i \in H^*$ corresponds to the element $t_{\alpha_i} \in H$. As H^* is spanned by the roots $\alpha_1, \ldots, \alpha_\ell$, H has as a basis $\{t_{\alpha_i} : 1 \leq i \leq \ell\}$. By Lemma 10.6, h_i is a non-zero scalar multiple of t_{α_i}. Hence $\{h_1, \ldots, h_\ell\}$ is a basis for H.

Now let $\beta \in \Phi$. We want to show that L_β is contained in the Lie subalgebra generated by the e_i and the f_i. Call this subalgebra $\tilde{L}$. By Proposition 11.14, we know that $\beta = w(\alpha_j)$, where w is a product of reflections s_{α_i} for some base elements α_i. Hence, by induction on the number of reflections, it is enough to prove the following: If $\beta = s_{\alpha_i}(\gamma)$ for some $\gamma \in \Phi$ with $L_\gamma \subseteq \tilde{L}$, then $L_\beta \subseteq \tilde{L}$.

By hypothesis, $\beta = \gamma - \langle \gamma, \alpha_i \rangle \alpha_i$. In Proposition 10.10, we looked at the $\mathsf{sl}(\alpha_i)$-submodule of L defined by

$$\bigoplus_k L_{\gamma + k\alpha_i},$$

where the sum is over all $k \in \mathbf{Z}$ such that $\gamma + k\alpha_i \in \Phi$, and the module structure is given by the adjoint action of $\mathsf{sl}(\alpha_i)$. We proved that this is an irreducible $\mathsf{sl}(\alpha_i)$-module. If $0 \neq e_\gamma \in L_\gamma$, then by applying powers of $\mathrm{ad}\, e$ or $\mathrm{ad}\, f$ we may obtain $e_{\gamma + k\alpha_i}$ whenever $\gamma + k\alpha_i \in \Phi$. Hence, if we take $k = \langle \gamma, \alpha_i \rangle$, then we will obtain e_β. Hence L_β is contained in $\tilde{L}$. $\qquad\square$

14.1.2 Relations

Next, we search for relations satisfied by the e_i, f_i, and h_i. These should only involve information which can be obtained from the Cartan matrix. We write $c_{ij} = \langle \alpha_i, \alpha_j \rangle$. Note that since the angle between any two base elements is obtuse (see Exercise 11.3), $c_{ij} \leq 0$ for all $i \neq j$.

Lemma 14.5

The elements e_i, f_i, h_i for $1 \leq i \leq \ell$ satisfy the following relations.

(S1) $[h_i, h_j] = 0$ for all i, j;

(S2) $[h_i, e_j] = c_{ji} e_j$ and $[h_i, f_j] = -c_{ji} f_j$ for all i, j;

(S3) $[e_i, f_i] = h_i$ for each i and $[e_i, f_j] = 0$ if $i \neq j$;

(S4) $(\mathrm{ad}\, e_i)^{1-c_{ji}}(e_j) = 0$ and $(\mathrm{ad}\, f_i)^{1-c_{ji}}(f_j) = 0$ if $i \neq j$.

Proof

We know H is a Cartan subalgebra and hence it is abelian, so (S1) holds. Condition (S2) follows from

$$[h_i, e_j] = \alpha_j(h_i)e_j = \langle \alpha_j, \alpha_i \rangle e_j = c_{ji} e_j,$$

while the first part of (S3) follows from the isomorphism of $\mathsf{sl}(\alpha_i)$ with $\mathsf{sl}(2, \mathbf{C})$. If $i \neq j$, we have $[e_i, f_j] \in L_{\alpha_i - \alpha_j}$; see Lemma 10.1(i). But since $\alpha_1, \ldots, \alpha_\ell$ form a base for Φ, $\alpha_i - \alpha_j \notin \Phi$. Therefore $L_{\alpha_i - \alpha_j} = 0$. This proves the second part of (S3).

To prove (S4), we fix α_i, α_j in the base and consider

$$M = \bigoplus_k L_{\alpha_j + k\alpha_i},$$

where the sum is taken over all $k \in \mathbf{Z}$ such that $\alpha_j + k\alpha_i \in \Phi$. As before, this is an $\mathsf{sl}(\alpha_i)$-module. Since $\alpha_j - \alpha_i \notin \Phi$, the sum only involves $k \geq 0$ and $k = 0$ does occur. Thus the smallest eigenvalue of $\operatorname{ad} h_i$ on M is $\langle \alpha_j, \alpha_i \rangle = c_{ji}$. By the classification of irreducible $\mathsf{sl}(2, \mathbf{C})$-modules in Chapter 8, the largest eigenvalue of $\operatorname{ad} h_i$ must be $-c_{ji}$.

An $\operatorname{ad} h_i$ eigenvector with eigenvalue $-c_{ji}$ is given by $x = (\operatorname{ad} e_i)^{-c_{ji}}(e_j)$, so applying $\operatorname{ad} e_i$ to x gives zero. This proves the first part of (S4). In fact, we have even proved that $1 - c_{ji}$ is the minimal integer $r \geq 0$ such that $(\operatorname{ad} e_i)^r(e_j) = 0$.

The other part of (S4) is proved by the same method. (Alternatively, one might note that the set of $-\alpha_j$ also is a base for the root system with standard basis $f_i, e_i, -h_i$.) $\square$

Serre's Theorem says that these relations completely determine the Lie algebra.

Theorem 14.6 (Serre's Theorem)

Let C be the Cartan matrix of a root system. Let L be the complex Lie algebra which is generated by elements e_i, f_i, h_i for $1 \leq i \leq \ell$, subject to the relations (S1) to (S4). Then L is finite-dimensional and semisimple with Cartan subalgebra H spanned by $\{h_1, \ldots, h_\ell\}$, and its root system has Cartan matrix C.

We immediately give our main application. Suppose that L is a complex semisimple Lie algebra with Cartan matrix C. By Lemma 14.5 this Lie algebra satisfies the Serre relations, so we can deduce that it must be isomorphic to the Lie algebra in Serre's Theorem with Cartan matrix C. Hence, up to isomorphism, there is just one Lie algebra for each root system. (We remarked at the end of Chapter 12 on some examples that support this statement.)

Serre's Theorem also solves the problem of constructing Lie algebras with the exceptional root systems G_2, F_4, E_6, E_7, and E_8: Just apply it with the Cartan matrix for the type required! Moreover, it shows that, up to isomorphism, there is just one exceptional Lie algebra for each type.

One might like to know whether the exceptional Lie algebras occur in any natural way. They had not been encountered until the classification. But subsequently, after looking for them, they have all been found as algebras of derivations of suitable algebras. See Exercise 14.4 below for an indication of the approaches used.

14.2 On the Proof of Serre's Theorem

We will now give an outline of the proof of Serre's Theorem. The full details are quite involved; they are given for example in Humphreys, *Introduction to Lie Algebras and Representation Theory*, [14].

Step 1. One first considers the Lie algebra $\mathcal{L}$ generated by the elements e_i, f_i, h_i for $1 \leq i \leq \ell$ which satisfies the relations (S1) to (S3) but where (S4) is not yet imposed. This Lie algebra is (usually) infinite-dimensional. Its structure had been determined before Serre by Chevalley, Harish-Chandra, and Jacobson.

One difficulty of studying $\mathcal{L}$ is that one cannot easily see how large it is, and therefore one needs some rather advanced technology: Just defining a Lie algebra by generators and relations may well produce something which is either much smaller or larger than one intended — see Exercise 14.2 for a small illustration of this.

The structure of $\mathcal{L}$ is as follows. Let $\mathcal{E}$ be the Lie subalgebra of $\mathcal{L}$ generated by $\{e_1, \ldots, e_\ell\}$, and let $\mathcal{F}$ be the Lie subalgebra of $\mathcal{L}$ generated by $\{f_1, \ldots, f_\ell\}$. Let H be the span of $\{h_1, \ldots, h_\ell\}$. Then, as a vector space,

$$\mathcal{L} = \mathcal{F} \oplus H \oplus \mathcal{E}.$$

We pause to give two examples.

Example 14.7

Consider the root system of type $A_1 \times A_1$ shown in Example 11.6(d). Here $\mathcal{E}$ is the Lie algebra generated by e_1 and e_2, with the only relations being those coming from the Jacobi identity and the anticommutativity of the Lie bracket. A Lie algebra of this kind is known as a *free Lie algebra* and, as long as it has at least two generators, it is infinite-dimensional.

If instead we take the root system of type A_1, then each of $\mathcal{E}$ and $\mathcal{F}$ is 1-dimensional and $\mathcal{L}$ is just $\mathsf{sl}(2, \mathbf{C})$. This is the only case where $\mathcal{L}$ is finite-dimensional.

Step 2. Now we impose the relations (S4) onto $\mathcal{L}$. Let $\mathcal{U}^+$ be the ideal of $\mathcal{E}$ generated by all θ_{ij}, where

$$\theta_{ij} := (\operatorname{ad} e_i)^{1-c_{ji}}(e_j).$$

Similarly, let $\mathcal{U}^-$ be the ideal of $\mathcal{F}$ generated by all θ_{ij}^-, where

$$\theta_{ij}^- := (\operatorname{ad} f_i)^{1-c_{ji}}(f_j).$$

Let $\mathcal{U} := \mathcal{U}^+ \oplus \mathcal{U}^-$, and let

$$N^+ := \mathcal{E}/\mathcal{U}^+, \quad N^- := \mathcal{F}/\mathcal{U}^-.$$

One shows that $\mathcal{U}^+$, $\mathcal{U}^-$, and hence $\mathcal{U}$ are actually ideals of $\mathcal{L}$. Hence the Lie algebra L in Serre's Theorem, which by definition is $\mathcal{L}/\mathcal{U}$, decomposes as

$$L = N^- \oplus H \oplus N^+.$$

By definition, $\mathcal{U}^+$ and $\mathcal{U}^-$ are invariant under $\operatorname{ad} h_i$ for each i and therefore $\operatorname{ad} h_i$ acts diagonally on L. One now has to show that L is finite-dimensional, with Cartan subalgebra H, and that the corresponding root space decomposition has a base giving the prescribed Cartan matrix.

Example 14.8

For the root system $A_1 \times A_1$, we have, by definition, $c_{12} = c_{21} = 0$ and hence $\mathcal{U}^+$ is the ideal generated by $(\operatorname{ad} e_1)(e_2)$ and $(\operatorname{ad} e_2)(e_1)$; that is, by $[e_1, e_2]$. This produces a very small quotient $\mathcal{E}/\mathcal{U}^+$, which is spanned by the cosets of e_1, e_2 and is 2-dimensional.

This is not quite obvious, so we sketch a proof. Given $x \in \mathcal{E}$, we can subtract off an element in the span of e_1 and e_2 to leave x as a sum of elements of the form $[u, v]$ for $u, v \in \mathcal{E}$. Now, as $[e_1, e_1] = [e_2, e_2] = 0$, the bracket $[e_1, e_2]$ must appear in every element in $\mathcal{E}'$ (when expressed in terms of e_1 and e_2), so $\mathcal{E}' = \mathcal{U}^+$ and $x \in \operatorname{Span}\{e_1, e_2\} + \mathcal{U}^+$.

Similarly, $\mathcal{F}/\mathcal{U}^-$ is 2-dimensional, spanned by the cosets of f_1 and f_2. Write $\bar{x}$ for the coset of x in L. We see directly that $L = \mathcal{L}/\mathcal{U}$ has a direct sum decomposition

$$\operatorname{Span}\{\bar{e}_1, \bar{f}_1, h_1\} \oplus \operatorname{Span}\{\bar{e}_2, \bar{f}_2, h_2\},$$

where $\bar{e}_1$ denote the coset $e + \mathcal{U}^+$, and so on. These are ideals in L, and each is isomorphic to $\mathsf{sl}(2, \mathbf{C})$, so in this case we get that L is the direct sum of two copies of $\mathsf{sl}(2, \mathbf{C})$, as we should expect.

For the general proof, more work is needed. The reader might like to try to construct a Lie algebra of type B_2 by this method to get some flavour of what is required.

14.3 Conclusion

The definition of a semisimple Lie algebra does not, on the face of it, seem very restrictive, so the fact that the complex semisimple Lie algebras are determined, up to isomorphism, by their Dynkin diagrams should seem quite remarkable.

In Appendix C, we show that the root system of a semisimple Lie algebra is uniquely determined (up to isomorphism). Thus complex semisimple Lie algebras with different Dynkin diagrams are not isomorphic. This is the last ingredient we need to establish a bijective correspondence between isomorphism classes of complex semisimple Lie algebras and the Dynkin diagrams listed in Theorem 13.1.

This classification theorem is one of the most important and far-reaching in mathematics; we look at some of the further developments it has motivated in the final chapter.

EXERCISES

14.1.   Use Serre's Theorem to show that the Lie algebra $\mathsf{so}(6, \mathbf{C})$ is isomorphic to $\mathsf{sl}(4, \mathbf{C})$. (This isomorphism can also be shown by geometric arguments; see Chapter 15.)

14.2.   Let L be the Lie algebra generated by x, y, z subject to the relations

$$[x, y] = z, \ [y, z] = x, \ [z, x] = x.$$

Show that L is one-dimensional.

14.3.   Let L be a Lie algebra generated by x, y, with no relations other than the Jacobi identity, and $[u, v] = -[v, u]$ for $u, v \in L$. Show that any Lie algebra $\mathcal{G}$ generated by two elements occurs as a homomorphic image of L. (So if you could establish that there are such $\mathcal{G}$ of arbitrary large dimensions, then you could deduce that L must be infinite-dimensional.)

14.4.   Let H be the algebra of quaternions. Thus H is the 4-dimensional real associative algebra with basis $1, i, j, k$ and multiplication described by $i^2 = j^2 = k^2 = ijk = -1$. (These are the equations famously carved in 1843 by Hamilton on Brougham Bridge in Dublin.)

(i) Let $\delta \in \mathrm{Der}\, H$, the Lie algebra of derivations of H. Show that δ preserves the subspace of H consisting of purely imaginary quaternions (that is, those elements of the form $xi+yj+zk$) and that $\delta(1) = 0$. Hence show that $\mathrm{Der}\, H$ is isomorphic to the

Lie algebra of antisymmetric 3×3 real matrices. (In particular, it has a faithful 3-dimensional representation.)

(ii) Show that if we complexify Der H by taking the algebra of antisymmetric 3×3 complex matrices, we obtain $\mathsf{sl}(2, \mathbf{C})$.

One step up from the quaternions lies the 8-dimensional Cayley algebra of octonions. One can construct the exceptional Lie algebra g_2 of type G_2 by taking the algebra of derivations of the octonions and then complexifying; this construction also gives its smallest faithful representation. The remaining exceptional Lie algebras can also be constructed by related techniques. For details, we refer the reader to either Schafer, *An Introduction to Nonassociative Algebras* [21] or Baez, "The Octonions"[2].

15
Further Directions

Now that we have a good understanding of the complex semisimple Lie algebras, we can also hope to understand their representation theory. This is the first of the topics we shall discuss. By Weyl's Theorem, every finite-dimensional representation is a direct sum of irreducible representations; we shall outline their construction. An essential tool in the representation theory of Lie algebras is the universal enveloping algebra associated to a Lie algebra. We explain what this is and why it is important.

The presentation of complex semisimple Lie algebras by generators and relations given in Serre's Theorem has inspired the definition of new families of Lie algebras. These include the Kac–Moody Lie algebras and their generalisations, which also have been important in the remarkable "moonshine" conjectures.

The theory of complex simple Lie algebras was used by Chevalley to construct simple groups of matrices over any field. The resulting groups are now known as *Chevalley groups* or as *groups of Lie type*. We briefly explain the basic idea and give an example.

Going in the other direction, given a group with a suitable 'smooth' structure, one can define an associated Lie algebra and use it to study the group. It was in fact in this way that Lie algebras were first discovered. We have given a very rough indication of this process in Exercise 4.9; as there are already many accessible books in this area, for example *Matrix Groups* by Baker [3] in the SUMS series, we refer the reader to them for further reading.

A very spectacular application of the theory of Lie algebra to group theory occurs in the restricted Burnside problem, which we discuss in §15.5. This involves Lie algebras defined over fields with prime characteristic. Lie algebras defined over fields of prime characteristic occur in several other contexts; we shall mention restricted Lie algebras and give an example of a simple Lie algebra that does not have an analogue in characteristic zero.

As well as classifying complex semisimple Lie algebras, Dynkin diagrams also appear in the representation theory of associative algebras. We shall explain some of the theory involved. Besides the appearance of Dynkin diagrams, one reason for introducing this topic is that there is a surprising connection with the theory of complex semisimple Lie algebras.

The survey in this chapter is certainly not exhaustive, and in places it is deliberately informal. Our purpose is to describe the main ideas; more detailed accounts exist and we give references to those that we believe would be accessible to the interested reader. For accounts of the early history of Lie algebras we recommend *Wilhelm Killing and the Structure of Lie algebras*, by Hawkins [12] and *The Mathematician Sophus Lie*, by Stubhaug [23].

15.1 The Irreducible Representations of a Semisimple Lie Algebra

We begin by describing the classification of the finite-dimensional irreducible representations of a complex semisimple Lie algebra L. By Weyl's Theorem, we may then obtain all finite-dimensional representations by taking direct sums of irreducible representations.

Let L have Cartan subalgebra H and root system Φ. Choose a base $\Pi = \{\alpha_1, \ldots, \alpha_\ell\}$ of Φ and let Φ^+ and Φ^- denote respectively the positive and negative roots with respect to Π. It will be convenient to use the *triangular decomposition*

$$L = N^- \oplus H \oplus N^+.$$

Here $N^+ = \bigoplus_{\alpha \in \Phi^+} L_\alpha$ and $N^- = \bigoplus_{\alpha \in \Phi^+} L_{-\alpha}$. Note that the summands H, N^-, and N^+ are subalgebras of L.

15.1.1 General Properties

Suppose that V is a finite-dimensional representation of L. Each element of H is semisimple, so it acts diagonalisably on L (see Exercise 9.14). Since finitely many commuting linear maps can be simultaneously diagonalised, V has a basis of simultaneous eigenvectors for H.

We can therefore decompose V into weight spaces for H. For $\lambda \in H^\star$, let

$$V_\lambda = \{v \in V : h \cdot v = \lambda(h)v \text{ for all } h \in H\}$$

and let Ψ be the set of $\lambda \in H^\star$ for which $V_\lambda \neq 0$. The *weight space decomposition* of V is then

$$V = \bigoplus_{\lambda \in \Psi} V_\lambda.$$

Example 15.1

(1) Let $V = L$ with the adjoint representation. Then weights are the same thing as roots, and the weight space decomposition is just the root space decomposition.

(2) Let $L = \mathsf{sl}(3, \mathbf{C})$, let H be the Cartan subalgebra of diagonal matrices, and let $V = \mathbf{C}^3$ be its natural representation. The weights that appear are $\varepsilon_1, \varepsilon_2, \varepsilon_3$, where $\varepsilon_i(h)$ is the i-th entry of the diagonal matrix h.

For each $\alpha \in \Phi$, we may regard V as a representation of $\mathsf{sl}(\alpha)$. In particular, this tells us that the eigenvalues of h_α acting on V are integers, and hence the weights in Ψ lie in the real span of the roots. We saw in §10.6 that this space is an inner-product space.

Example 15.2

For example, the following diagram shows the weights of the natural and adjoint representations of $\mathsf{sl}(3, \mathbf{C})$ with respect to the Cartan subalgebra H of diagonal matrices projected onto a plane. The weight spaces of the natural representation are marked. To locate ε_1 we note that restricted to H, $\varepsilon_1 + \varepsilon_2 + \varepsilon_3 = 0$, and hence ε_1 is the same map on H as $\frac{1}{3}(2\varepsilon_1 - \varepsilon_2 - \varepsilon_3) = \frac{1}{3}(2\alpha + \beta)$.

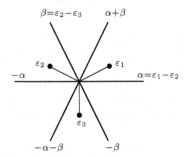

We now look at the action of e_α and f_α for $\alpha \in \Phi$. Let $v \in V_\lambda$. We leave it to the reader to check that $e_\alpha \cdot v \in V_{\lambda+\alpha}$ and $f_\alpha \cdot v \in V_{\lambda-\alpha}$; note that this generalises Lemma 8.3 for $\mathsf{sl}(2,\mathbf{C})$. Since the set Ψ of weights of V is finite, there must be some $\lambda \in \Psi$ such that for all $\alpha \in \Phi^+$, $\lambda + \alpha \notin \Psi$. We call such a λ a *highest weight*, and if $v \in V_\lambda$ is non-zero, then we say v is a highest-weight vector.

This agrees with our usage of these words in Chapter 8 since for representations of $\mathsf{sl}(2,\mathbf{C})$ a weight of the Cartan subalgebra spanned by h is essentially the same thing as an eigenvalue of h.

In the previous example, the positive roots of $\mathsf{sl}(3,\mathbf{C})$ with respect to the base $\Pi = \{\alpha,\beta\}$ are $\alpha, \beta, \alpha+\beta$, and so the (unique) highest weight of the natural representation of $\mathsf{sl}(3,\mathbf{C})$ is ε_1.

Lemma 15.3

Let V be a simple L-module. The set Ψ of weights of V contains a unique highest weight. If λ is this highest weight then V_λ is 1-dimensional and all other weights of V are of the form $\lambda - \sum_{\alpha_i \in \Pi} a_i \alpha_i$ for some $a_i \in \mathbf{Z}$, $a_i \geq 0$.

Proof

Take $0 \neq v \in V_\lambda$ and let W be the subspace of V spanned by elements of the form

$$f_{\alpha_{i_1}} f_{\alpha_{i_2}} \cdots f_{\alpha_{i_k}} \cdot v, \qquad (\star)$$

where the α_{i_j} are not necessarily distinct elements of Π. Note that each element of the form $(\star)$ is an H-eigenvector. We claim that W is an L-submodule of V.

By Lemma 14.4, L is generated by the elements e_α, f_α for $\alpha \in \Pi$, so it is enough to check that W is closed under their action. For the f_α, this follows at once from the definition. Let $w = f_{\alpha_{i_1}} f_{\alpha_{i_2}} \cdots f_{\alpha_{i_k}} \cdot v$. To show that $e_\alpha \cdot w \in W$, we shall use induction on k.

If $k = 0$ (that is, $w = v$), then we know that $e_\alpha \cdot v = 0$. For $k \geq 1$, let $w_1 = f_{\alpha_{i_2}} \cdots f_{\alpha_{i_k}} v$ so that $w = f_{\alpha_{i_1}} w_1$ and

$$e_\alpha \cdot w = e_\alpha \cdot (f_{\alpha_{i_1}} \cdot w_1) = f_{\alpha_{i_1}} \cdot (e_\alpha \cdot w_1) + [e_\alpha, f_{\alpha_{i_1}}] \cdot w_1.$$

Now $[e_\alpha, f_{\alpha_{i_1}}] \in [L_\alpha, L_{-\alpha_{i_1}}] \subseteq L_{\alpha - \alpha_{i_1}}$. Both α and α_{i_1} are elements of the base Π, so $L_{\alpha - \alpha_{i_1}} = 0$, unless $\alpha = \alpha_{i_1}$, in which case $L_{\alpha - \alpha_{i_1}} \subseteq L_0 = H$. So in either case w_1 is an eigenvector for $[f_{\alpha_{i_1}}, e_\alpha]$. Moreover, by the inductive hypothesis, $e_\alpha \cdot w_1$ lies in W, so by the definition of W we have $f_{\alpha_{i_1}} \cdot (e_\alpha \cdot w_1) \in W$.

Since V is simple and W is non-zero, we have $V = W$. We can see from $(\star)$ that the weights of V are of the form $\lambda - \sum_i a_i \alpha_i$ for $\alpha_i \in \Pi$ and $a_i \geq 0$, so λ is the unique highest weight. $\qquad\square$

Example 15.4

Let $L = \mathsf{sl}(\ell + 1, \mathbf{C})$ and let $V = L$, with the adjoint representation. By Example 7.4, V is a simple L-module. We have seen above that the root space decomposition of L is the same as the weight space decomposition of this module. The unique highest weight is $\alpha_1 + \alpha_2 + \ldots + \alpha_\ell$, and for the highest-weight vector v in the lemma we can take $e_{1,\ell+1}$.

Suppose that λ is a weight for a finite-dimensional representation V. Let $\alpha \in \Pi$. Suppose that $\lambda(h_\alpha)$, the eigenvalue of h_α on the λ-weight space, is negative. Then, by the representation theory of $\mathsf{sl}(2, \mathbf{C})$, $e_\alpha \cdot L_\lambda \neq 0$, and so $\alpha + \lambda \in \Psi$. Thus, if λ is the highest weight for a finite-dimensional representation V, then $\lambda(h_\alpha) \geq 0$ for all $\alpha \in \Pi$.

This motivates the main result, given in the following theorem.

Theorem 15.5

Let Λ be the set of all $\lambda \in H^*$ such that $\lambda(h_\alpha) \in \mathbf{Z}$ and $\lambda(h_\alpha) \geq 0$ for all $\alpha \in \Pi$. For each $\lambda \in \Lambda$, there is a finite-dimensional simple L-module, denoted by $V(\lambda)$, which has highest weight λ. Moreover, any two simple L-modules with the same highest weight are isomorphic, and every simple L-module may be constructed in this way.

To describe Λ in general, one uses the *fundamental dominant weights*. These are defined to be the unique elements $\lambda_1, \ldots, \lambda_\ell \in H^*$ such that

$$\lambda_i(h_{\alpha_j}) = \delta_{ij}.$$

By the theorem above, Λ is precisely the set of linear combinations of the λ_i with non-negative integer coefficients. One would also like to relate the λ_i to the elements of our base of H^*. Recall that $\lambda(h_\alpha) = \langle \lambda, \alpha \rangle$; so if we write $\lambda_i = \sum_{k=1}^{\ell} d_{ik}\alpha_k$, then

$$\lambda_i(h_{\alpha_j}) = \sum_k d_{ik}\langle \alpha_k, \alpha_j \rangle,$$

so the coefficients d_{ik} are given by the inverse of the Cartan matrix of L.

Example 15.6

Let $L = \mathsf{sl}(3, \mathbf{C})$. Then the inverse of the Cartan matrix is

$$\frac{1}{3} \begin{pmatrix} 2 & 1 \\ 1 & 2 \end{pmatrix},$$

and the fundamental dominant weights are $\frac{1}{3}(2\alpha + \beta) = \varepsilon_1$ and $\frac{1}{3}(\alpha + 2\beta) = -\varepsilon_3$. The diagram in 15.2 shows that ε_1 is the highest weight of the natural representation; ε_3 appears as the highest weight of the dual of the natural representation.

So far, for a general complex simple Lie algebra L, the only irreducible representations we know are the trivial and adjoint representations. If L is a classical Lie algebra, then we can add the natural representation to this list. The previous theorem says there are many more representations. How can they be constructed?

15.1.2 Exterior Powers

Several general methods of constructing new modules from old ones are known. Important amongst these are tensor products and the related symmetric and exterior powers.

Let V be a finite-dimensional complex vector space with basis $v_1, \ldots, v_n$. For each i, j with $1 \leq i, j \leq n$, we introduce a symbol $v_i \wedge v_j$, which satisfies $v_j \wedge v_i = -v_i \wedge v_j$. The *exterior square* $V \bigwedge V$ is defined to be the complex vector space of dimension $\binom{n}{2}$ with basis given by $\{v_i \wedge v_j : 1 \leq i < j \leq n\}$. Thus, a general element of $V \bigwedge V$ has the form

$$\sum_{i<j} c_{ij} v_i \wedge v_j \quad \text{for scalars } c_{ij} \in \mathbf{C}.$$

For $v = \sum a_i v_i$ and $w = \sum b_j v_j$, define $v \wedge w$ by

$$v \wedge w = \sum_{i,j} a_i b_j v_i \wedge v_j.$$

This shows that the map $(v, w) \to v \wedge w$ is bilinear. One can show that the definition does not depend on the choice of basis. That is, if $w_1, \ldots, w_n$ is some other basis of V, then the set of all $w_i \wedge w_j$ for $1 \leq i < j \leq n$ is a basis for $V \bigwedge V$ with the same properties as the previous basis.

Now suppose that L is a Lie algebra and $\rho : L \to \mathsf{gl}(V)$ is a representation. We may define a new representation $\wedge^2 \rho : L \to \mathsf{gl}(V \bigwedge V)$ by

$$(\wedge^2 \rho)(x)(v_i \wedge v_j) = \rho(x)v_i \wedge v_j + v_i \wedge \rho(x)v_j \quad \text{for } x \in L$$

and extending it to linear combinations of basis elements. (The reader might care to check that this really does define a representation of L.)

More generally, for any integer $r \leq n$, one introduces similarly symbols $v_{i_1} \wedge v_{i_2} \wedge \ldots \wedge v_{i_r}$ satisfying

$$v_{i_1} \wedge \ldots \wedge v_{i_k} \wedge v_{i_{k+1}} \wedge \ldots \wedge v_{i_r} = -v_{i_1} \wedge \ldots \wedge v_{i_{k+1}} \wedge v_{i_k} \wedge \ldots \wedge v_{i_r}.$$

The r-fold exterior power of V, denoted by $\bigwedge^r V$ is the vector space over $\mathbf{C}$ of dimension $\binom{n}{r}$ with basis

$$v_{i_1} \wedge v_{i_2} \wedge \ldots \wedge v_{i_r}, \quad 1 \leq i_1 < \ldots < i_r \leq n.$$

The action of L generalises so that

$$(\wedge^r \rho)(x)(v_{i_1} \wedge \ldots \wedge v_{i_r}) = \sum_{s=1}^{r} v_{i_1} \wedge \ldots \wedge \rho(x)v_{i_s} \wedge \ldots \wedge v_{i_r}.$$

It is known that if V is the natural module of a classical Lie algebra, then all the exterior powers V are irreducible. This is very helpful when constructing the irreducible representations of the classical Lie algebras.

We shall now use exterior powers to give a direct proof that $\mathsf{so}(6, \mathbf{C})$ and $\mathsf{sl}(4, \mathbf{C})$ are isomorphic. (In Chapter 14, we noted that this follows from Serre's Theorem, but we did not give an explicit isomorphism.)

Let $L = \mathsf{sl}(4, \mathbf{C})$, and let V be the 4-dimensional natural L-module. Then $\bigwedge^2 V$ is a 6-dimensional L-module. Now $\bigwedge^4 V$ has dimension $\binom{4}{4} = 1$. If we fix a basis $v_1, \ldots, v_4$ of V, then $\bigwedge^4 V$ is spanned by $\tilde{v} := v_1 \wedge v_2 \wedge v_3 \wedge v_4$. We may define a bilinear map

$$\bigwedge^2 V \times \bigwedge^2 V \to \mathbf{C}$$

by setting $(v, w) = c$ if $v \wedge w = c\tilde{v}$ for $c \in \mathbf{C}$.

Exercise 15.1

Find the matrix describing this bilinear form on $\bigwedge^2 V$ with respect to the basis $\{v_i \wedge v_j : i < j\}$. Show that it is congruent to the bilinear form defined by the matrix S, where

$$S = \begin{pmatrix} 0 & I \\ I & 0 \end{pmatrix}.$$

The module $\bigwedge^4 V$ is a 1-dimensional module for a semisimple Lie algebra, so it must be the trivial module for L. So for $x \in L$ and $v, w \in \bigwedge^2 V$, we have $x \cdot (v \wedge w) = 0$. But by the definition of the action of L, we get

$$x \cdot (v \wedge w) = v \wedge (xw) + (xv) \wedge w.$$

Hence, if we translate this into the bilinear form, we have

$$(v, xw) = -(xv, w).$$

Thus the image of $\varphi : L \to \mathsf{gl}(\bigwedge^2 V)$ is contained in $\mathsf{gl}_S(6, \mathbf{C}) = \mathsf{so}(6, \mathbf{C})$, where S is as above. Since L is simple and φ is non-zero, φ must be one-to-one, so by dimension counting it gives an isomorphism between $\mathsf{sl}(4, \mathbf{C})$ and $\mathsf{so}(6, \mathbf{C})$.

15.1.3 Tensor Products

Let V and W be finite-dimensional complex vector spaces with bases $v_1, \ldots, v_m$ and $w_1, \ldots, w_n$, respectively. For each i, j with $1 \le i \le m$ and $1 \le j \le n$, we introduce a symbol $v_i \otimes w_j$. The tensor product space $V \otimes W$ is defined to be the mn-dimensional complex vector space with basis given by $\{v_i \otimes w_j : 1 \le i \le n, 1 \le j \le m\}$. Thus a general element of $V \otimes W$ has the form

$$\sum_{i,j} c_{ij} v_i \otimes w_j \quad \text{for scalars } c_{ij} \in \mathbf{C}.$$

For $v = \sum_i a_i v_i \in V$ and $w = \sum_j b_j w_j \in W$, we define $v \otimes w \in V \otimes W$ by

$$v \otimes w = \sum_{i,j} a_i b_j (v_i \otimes w_j).$$

This shows that $(v, w) \to v \otimes w$ is bilinear. Again one can show that this definition of $V \otimes W$ does not depend on the choice of bases.

Suppose we have representations $\rho_1 : L \to \mathsf{gl}(V)$, $\rho_2 : L \to \mathsf{gl}(W)$. We may define a new representation $\rho : L \to \mathsf{gl}(V \otimes W)$ by

$$\rho(x)(v \otimes w) = \rho_1(x)(v) \otimes w + v \otimes \rho_2(x)(w).$$

Example 15.7

Let $L = \mathsf{sl}(2, \mathbf{C})$, and let $V = \mathbf{C}^2$ be the natural module with standard basis v_1, v_2. Let $W = \mathbf{C}^2$ be another copy of the natural module, with basis w_1, w_2. With respect to the basis $v_1 \otimes v_1, v_1 \otimes v_2, v_2 \otimes v_1, v_2 \otimes v_2$, one finds that the matrices of e, f, and h are

$$\rho(e) = \begin{pmatrix} 0 & 1 & 1 & 0 \\ 0 & 0 & 0 & 1 \\ 0 & 0 & 0 & 1 \\ 0 & 0 & 0 & 0 \end{pmatrix}, \ \rho(f) = \begin{pmatrix} 0 & 0 & 0 & 0 \\ 1 & 0 & 0 & 0 \\ 1 & 0 & 0 & 0 \\ 0 & 1 & 1 & 0 \end{pmatrix}, \ \rho(h) = \begin{pmatrix} 2 & 0 & 0 & 0 \\ 0 & 0 & 0 & 0 \\ 0 & 0 & 0 & 0 \\ 0 & 0 & 0 & -2 \end{pmatrix}.$$

By Exercise 8.4, an $\mathsf{sl}(2, \mathbf{C})$-module is determined up to isomorphism by the eigenvalues of h. Here the highest eigenvalue appearing is 2, so V_2 is a submodule of $V \otimes W$. This leaves only an eigenvalue of 0, so we must have

$$V \otimes W \cong V_0 \oplus V_2.$$

Exercise 15.2

Find an explicit direct sum decomposition of $V \otimes W$ into irreducible submodules.

The book *Representation Theory* by Fulton and Harris [10] works out many more examples of this type.

For a general semisimple Lie algebra L, there is a more efficient and unified construction of the simple L-modules, which also allows one to construct certain infinite-dimensional representations. This uses the universal enveloping algebra of L. We shall now introduce this algebra and explain how to use it to construct the simple L-modules. We also explain the main idea in the proof of Theorem 15.5 above.

15.2 Universal Enveloping Algebras

Given a Lie algebra L over a field F, one can define its *universal enveloping algebra*, denoted by $U(L)$. This is an associative algebra (see §1.5) over F, which is always infinite-dimensional unless L is zero.

Assume that L is finite-dimensional with vector space basis $\{x_1, x_2, \ldots, x_n\}$. The structure constants with respect to this basis are the scalars a_{ij}^k given by

$$[x_i, x_j] = \sum_k a_{ij}^k x_k \quad \text{for } 1 \le i, j \le n.$$

Then $U(L)$ can be defined as the unital associative algebra, generated by $X_1, X_2, \ldots, X_n$, subject to the relations

$$X_i X_j - X_j X_i = \sum_{k=1}^{n} a_{ij}^k X_k \quad \text{for } 1 \le i, j \le n.$$

It can be shown (see Exercise 15.8) that the algebra $U(L)$ does not depend on the choice of the basis. That is, if we start with two different bases for L, then the algebras we get by this construction are isomorphic.

Example 15.8

(1) Let $L = \text{Span}\{x\}$ be a 1-dimensional abelian Lie algebra over a field F. The only structure constants come from $[x, x] = 0$. This gives us the relation $XX - XX = 0$, which is vacuous. Hence $U(L)$ is the associative algebra generated by the single element X. In other words, $U(L)$ is the polynomial algebra $F[X]$.

(2) More generally, let L be the n-dimensional abelian Lie algebra with basis $\{x_1, x_2, \ldots, x_n\}$. As before, all structure constants are zero, and hence $U(L)$ is isomorphic to the polynomial algebra in n variables.

We now consider a more substantial example. Let $L = \mathsf{sl}(2, \mathbf{C})$ with its usual basis, f, h, e. We know the structure constants and therefore we can calculate in the algebra $U(L)$. We should really write F, H, E for the corresponding generators of $U(L)$, but unfortunately this creates an ambiguity as H is already used to denote Cartan subalgebras. So instead we also write f, h, e for the generators of $U(L)$; the context will make clear the algebra in which we are working.

The triangular decomposition of L,

$$L = N^- \oplus H \oplus N^+,$$

where $N^- = \mathrm{Span}\{f\}$, $H = \mathrm{Span}\{h\}$, and $N^+ = \mathrm{Span}\{e\}$, gives us three subalgebras of $U(L)$. For example, $U(L)$ contains all polynomials in e; this subalgebra can be thought of as the universal enveloping algebra $U(N^+)$. Similarly, $U(L)$ contains all polynomials in f and in h. But, in addition, $U(L)$ contains products of these elements. Using the relations $ef - fe = h$, $he - eh = 2e$, and $hf - fh = -2f$, valid in $U(L)$, one can show the following.

Lemma 15.9

Let $L = \mathsf{sl}(2, \mathbf{C})$. The associative algebra $U(L)$ has as a vector space basis

$$\{f^a h^b e^c : a, b, c \geq 0\}.$$

To show that this set spans the universal enveloping algebra, it suffices to verify that every monomial in the generators can be expressed as a linear combination of monomials of the type appearing in the lemma. The reader might, as an exercise, express the monomial hef as a linear combination of the given set; this should be enough to show the general strategy.

Proving linear independence is considerably harder, so we shall not go into the details. Indeed it is not even obvious that the elements $e, f \in U(L)$ are linearly independent, but this much at least will follow from Exercise 15.8.

In general, if the Lie algebra L has basis $x_1, \ldots, x_n$, then the algebra $U(L)$ has basis

$$\{X_1^{a_1} X_2^{a_2} \ldots X_n^{a_n} : a_1, \ldots a_n \geq 0\}.$$

This is known as a Poincaré–Birkhoff–Witt-basis or PBW-basis of $U(L)$. The previous lemma is the special case where $L = \mathsf{sl}(2, \mathbf{C})$ and $X_1 = f$, $X_2 = h$, and $X_3 = e$. We could equally well have taken the basis elements in a different order.

An important corollary is that the elements X_1, X_2, ..., X_n are linearly independent, and so L can be found as a subspace of $U(L)$. Furthermore, if L_1 is a Lie subalgebra of L, then $U(L_1)$ is an associative subalgebra of $U(L)$; this justifies our earlier assertions about polynomial subalgebras of $U(\mathsf{sl}(2, \mathbf{C}))$.

15.2.1 Modules for $U(L)$

We now explain the sense in which the universal enveloping algebra of a Lie algebra L is "universal". We first need to introduce the idea of a representation of an associative algebra.

Let A be a unital associative algebra over a field F. A *representation* of A on an F-vector space V is a homomorphism of associative algebras

$$\varphi : A \to \mathrm{End}_F(V),$$

where $\mathrm{End}_F(V)$ is the associative algebra of linear maps on V. Thus φ is linear, φ maps the multiplicative identity of A to the identity map of V, and

$$\varphi(ab) = \varphi(a) \circ \varphi(b) \quad \text{for all } a, b \in A.$$

Unlike in earlier chapters, we now allow V to be infinite-dimensional. Note that the underlying vector space of $\mathrm{End}_F(V)$ is the same as that of $\mathsf{gl}(V)$; we write $\mathrm{End}_F(V)$ if we are using its associative structure and $\mathsf{gl}(V)$ if we are using its Lie algebra structure.

In what follows, it is most convenient to use the language of modules, so we shall indicate the action of L implicitly by writing $a \cdot v$ rather than $\varphi(a)(v)$.

Lemma 15.10

Let L be a Lie algebra and let $U(L)$ be its universal enveloping algebra. There is a bijective correspondence between L-modules and $U(L)$-modules. Under this correspondence, an L-module is simple if and only if it is simple as a module for $U(L)$.

Proof

Let V be an L-module. Since the elements X_i generate $U(L)$ as an associative algebra, the action $U(L)$ on V is determined by the action of the X_i. We let $X_i \in U(L)$ act on V in the same way as $x_i \in L$ acts on V. To verify that this defines an action of $U(L)$, one only needs to check that it satisfies the defining relations for $U(L)$. Consider the identity in L

$$[x_i, x_j] = \sum_k a_{ij}^k x_k.$$

For the action to be well-defined, we require that on V

$$(X_i X_j - X_j X_i)v = \sum_k a_{ij}^k X_k v.$$

By definition, the left-hand side is equal to $(x_i x_j - x_j x_i)v$; that is,

$$[x_i, x_j]v = \sum_k a_{ij}^k x_k v.$$

Since X_k acts on V in the same way as x_k, this is equal to the right-hand side, as we required.

Conversely, suppose V is a $U(L)$-module. By restriction, V is also an L-module since $L \subseteq U(L)$. Furthermore, V is simple as an L-module if and only if it is simple as a module for $U(L)$. This is a simple change of perspective and can easily be checked formally. $\square$

The proof of this lemma demonstrates a certain *universal property* of $U(L)$. See Exercise 15.8 for more details.

15.2.2 Verma Modules

Suppose that L is a complex semisimple Lie algebra and $U(L)$ is the universal enveloping algebra of L. We shall use the equivalence between modules for $U(L)$ and L to construct an important family of L-modules.

Let H be a Cartan subalgebra of L, let Φ be the corresponding root system, and let Π be a base of Φ. As usual, we write Φ^+ for the positive roots with respect to Π. We may choose a basis $h_1, \ldots, h_\ell$ of H such that $h_i = h_{\alpha_i}$ for $\alpha_i \in \Pi$. For $\lambda \in H^*$ let $I(\lambda)$ be the left ideal of $U(L)$ generated by the elements e_α for $\alpha \in \Phi$ and also $h_i - \lambda(h_i)1$ for $1 \leq i \leq \ell$. Thus $I(\lambda)$ consists of all elements

$$\sum u_\alpha e_\alpha + \sum y_i (h_i - \lambda(h_i)1),$$

where the u_α and the y_i are arbitrary elements of $U(L)$. We may consider $I(\lambda)$ as a left module for $U(L)$. Let $M(\lambda)$ be the quotient space

$$M(\lambda) := U(L)/I(\lambda).$$

This becomes a $U(L)$-module with the action $u \cdot (v + I(\lambda)) = uv + I(\lambda)$. We say $M(\lambda)$ is the *Verma module* associated to λ.

Proposition 15.11

If $\bar{v} = 1 + I(\lambda)$, then $\bar{v}$ generates $M(\lambda)$ as a $U(L)$-module. For $\alpha \in \Phi^+$ and $e_\alpha \in L_\alpha$, we have $e_\alpha \bar{v} = 0$; and for $h \in H$ we have $h\bar{v} = \lambda(h)\bar{v}$. The module $M(\lambda)$ has a unique maximal submodule, and the quotient of $M(\lambda)$ by this submodule is the simple module $V(\lambda)$ with highest weight λ.

The first part of the theorem is easy: We have

$$e_\alpha \cdot \bar{v} = e_\alpha + I(\lambda),$$

which is zero in $M(\lambda)$. Moreover,

$$h_i \cdot \bar{v} = h_i + I(\lambda) = \lambda(h_i)1 + I(\lambda)$$

since $h_i - \lambda(h_i)1 \in I(\lambda)$. Since

$$x + I(\lambda) = x \cdot (1 + I(\lambda)) = x \cdot \bar{v} \quad \text{for all } x \in U(L),$$

the coset $\bar{v}$ generates $M(\lambda)$.

One can show that a vector space basis for $M(\lambda)$ is given by the elements $u \cdot \bar{v}$, where u runs through a basis of $U(N^-)$. By the PBW-Theorem, $U(N^-)$ has a basis consisting of monomials in the f_α for $\alpha \in \Phi$; this shows that $M(\lambda)$ decomposes as a direct sum of simultaneous H-eigenspaces. We can then see that $M(\lambda)$ has a unique maximal weight, namely λ.

Knowing this, one can complete the proof of the proposition. Details can be found in Humphreys [14] (Chapter 20), or Dixmier [9]. Note, however, that the labelling in Dixmier is slightly different.

Example 15.12

We give two examples of Verma modules for $L = \mathsf{sl}(2, \mathbf{C})$. First we construct one which is irreducible; this will show that L has infinite-dimensional irreducible representations.

(1) Let $\lambda = -d$, where $d > 0$. Thus $M(\lambda) = U(L)/I(\lambda)$, where

$$I(\lambda) = U(L)e + U(L)(h + d1).$$

As a vector space, $M(\lambda)$ has basis

$$\left\{ \bar{f}^a = f^a + I_\lambda : a \geq 0 \right\}.$$

It follows by induction for each $a \geq 0$ that $\bar{f}^a$ is an eigenvector for h with eigenvalue $-d - 2a$. Furthermore, we have $e \cdot \bar{1} = 0$, $e \cdot \bar{f} = -d \cdot \bar{1}$, and inductively $e \cdot \bar{f}^a = c_{a-1} \bar{f}^{a-1}$, where c_{a-1} is a negative integer. As in §8.1.1, we can draw $M(\lambda)$ as

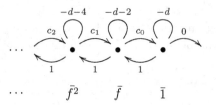

where loops represent the action of h, arrows to the right represent the action of e, and arrows to the left represent the action of f. Using this, one can check that for any non-zero $x \in M(\lambda)$ the span of

$$\{x, e \cdot x, e^2 \cdot x, \ldots\}$$

contains the generator $\bar{1} = 1 + I(\lambda)$ and hence $M(\lambda)$ is an infinite-dimensional simple module.

(2) We consider the Verma module $M(0)$. In this case, the span of all $\bar{f}^a$ where $a > 0$ is a proper submodule of $M(\lambda)$. For example,

$$e \cdot \bar{f} = ef + I(0) = (fe + h) + I(0),$$

which is zero, since $e, h \in I(0)$. The quotient of $M(0)$ by this submodule is the trivial L-module, $V(0)$.

Verma modules are the building blocks for the so-called *category* $\mathcal{O}$, which has recently been of major interest. Here the starting point is the observation that although each $M(\lambda)$ is infinite-dimensional, when viewed as $U(L)$-module it has finite length. That is, there are submodules

$$0 = M_0 \subseteq M_1 \subseteq M_2 \subseteq \ldots \subseteq M_k = M(\lambda)$$

such that M_i/M_{i-1} is simple for $1 \leq i \leq k$. A proof of this and more properties of Verma modules can be found in Dixmier [9] or Humphreys [14] (note, however, that Dixmier uses different labelling.)

In 1985, Drinfeld and Jimbo independently defined *quantum groups* by "deforming" the universal enveloping algebras of Lie algebras. (So contrary to what one might expect, quantum groups are really algebras!) Since then, quantum groups have found numerous applications in areas including theoretical physics, knot theory, and representations of algebraic groups. In 1990, Drinfeld was awarded a Fields Medal for his work. For more about quantum groups, see Jantzen, *Lectures on Quantum Groups* [16].

15.3 Groups of Lie Type

The theory of simple Lie algebras over $\mathbf{C}$ was used by Chevalley to construct simple groups of matrices over any field.

How can one construct invertible linear transformations from a complex Lie algebra? Let $\delta : L \to L$ be a derivation of L such that $\delta^n = 0$ for some $n \geq 1$. In Exercise 15.9, we define the exponential $\exp(\delta)$ and show that it is an automorphism of L.

Given a complex semisimple Lie algebra L, let x be an element in a root space. We know that $\operatorname{ad} x$ is a derivation of L, and by Exercise 10.1 $\operatorname{ad} x$ is nilpotent. Hence $\exp(\operatorname{ad} x)$ is an automorphism of L. One then takes the group generated by all the $\exp(\operatorname{ad} cx)$, for $c \in \mathbf{C}$, for x in a strategically chosen basis of L. This basis is known as the *Chevalley basis*; it is described in the following theorem.

Theorem 15.13

Let L be a simple Lie algebra over $\mathbf{C}$, with Cartan subalgebra H and associated root system Φ, and let Π be a base for Φ. For each $\alpha \in \Phi$, one may choose $h_\alpha \in H$ so that $h_\alpha \in [L_{-\alpha}, L_\alpha]$ and $\alpha(h_\alpha) = 2$. One may also choose an element $e_\alpha \in L_\alpha$ such that $[e_\alpha, e_{-\alpha}] = h_\alpha$ and $[e_\alpha, e_\beta] = \pm(p+1)e_{\alpha+\beta}$, where p is the greatest integer for which $\beta + p\alpha \in \Phi$.

The set $\{h_\alpha : \alpha \in \Pi\} \cup \{e_\beta : \beta \in \Phi\}$ is a basis for L. Moreover, for all $\gamma \in \Phi$, $[e_\gamma, e_{-\gamma}] = h_\gamma$ is an integral linear combination of the h_α for $\alpha \in \Pi$. The remaining structure constants of L with respect to this basis are as follows:

$$[h_\alpha, h_\beta] = 0,$$
$$[h_\alpha, e_\beta] = \beta(h_\alpha)e_\beta,$$
$$[e_\alpha, e_\beta] = \begin{cases} \pm(p+1)e_{\alpha+\beta} & \alpha + \beta \in \Phi \\ 0 & \alpha + \beta \notin \Phi \cup \{0\}. \end{cases}$$

In particular, they are all integers. $\qquad\qquad\square$

Recall that in §10.4 we found for each $\alpha \in \Phi$ a subalgebra $\operatorname{Span}\{e_\alpha, f_\alpha, h_\alpha\}$ isomorphic to $\mathsf{sl}(2, \mathbf{C})$. Chevalley's Theorem asserts that the e_α and $f_\alpha = e_{-\alpha}$ can be chosen so as to give an especially convenient form for the structure constants of L.

Exercise 15.3

By using the calculations in Chapter 12, determine a Chevalley basis for the Lie algebra $\mathsf{so}(5, \mathbf{C})$ of type B_2.

Since the structure constants are integers, the $\mathbf{Z}$-span of such a basis, denoted by $L_{\mathbf{Z}}$, is closed under Lie brackets. If one now takes any field F, one can define a Lie algebra L_F over F as follows. Take as a basis

$$\{\bar{h}_\alpha : \alpha \in \Pi\} \cup \{\bar{e}_\beta, \beta \in \Phi\}$$

and define the Lie commutator by taking the structure constants for $L_{\mathbf{Z}}$ and interpreting them as elements in the prime subfield of F. For example, the standard basis e, h, f of $\mathsf{sl}(2, \mathbf{C})$ is a Chevalley basis, and applying this construction gives $\mathsf{sl}(2, F)$.

Now we can describe the automorphisms. First take the field $\mathbf{C}$. For $c \in \mathbf{C}$ and $\alpha \in \Phi$, define

$$x_\alpha(c) := \exp(c \operatorname{ad} e_\alpha).$$

As explained, this is an automorphism of L. One can show that it takes elements of the Chevalley basis to linear combinations of basis elements with coefficients of the form ac^i, where $a \in \mathbf{Z}$ and $i \geq 0$. Let $A_\alpha(c)$ be the matrix of $x_\alpha(c)$ with respect to the Chevalley basis of L. By this remark, the entries of $A_\alpha(c)$ have the form ac^i for $a \in \mathbf{Z}$ and $i \geq 0$. Define the *Chevalley group* associated to L by

$$G_{\mathbf{C}}(L) := \langle A_\alpha(c) : \alpha \in \Phi, c \in \mathbf{C} \rangle.$$

We can also define automorphisms of L_F. Take $t \in F$. Let $\tilde{A}_\alpha(t)$ be the matrix obtained from $A_\alpha(c)$ by replacing each entry ac^i by $\bar{a}t^i$, where $\bar{a}$ is a viewed as an element in the prime subfield of F. The Chevalley group of L over F is then defined to be the group

$$G_F(L) := \langle \tilde{A}_\alpha(t) : \alpha \in \Phi, t \in F \rangle.$$

Exercise 15.4

Let $L = \mathsf{sl}(2, \mathbf{C})$. Let $c \in \mathbf{C}$. Show that with respect to the Chevalley basis e, h, f, the matrix of $\exp(c \operatorname{ad} e)$ is

$$\begin{pmatrix} 1 & -2c & -c^2 \\ 0 & 1 & c \\ 0 & 0 & 1 \end{pmatrix}$$

and find the matrix of $\exp(c \operatorname{ad} f)$. Then describe the group $G_{\mathbf{F}_2}(L)$, where $\mathbf{F}_2$ is the field with 2-elements.

The structure of these groups is studied in detail in Carter's book *Simple Groups of Lie Type* [7], see also [13].

Remark 15.14

One reason why finite groups of Lie type are important is the Classification Theorem of Finite Simple Groups. This theorem, which is one of the greatest achievements of twentieth century mathematics (though to date not yet completely written down), asserts that there are two infinite families of finite simple groups, namely the alternating groups and the finite groups of Lie type, and that any finite simple group is either a member of one of these two families or is one of the 26 sporadic simple groups.

15.4 Kac–Moody Lie Algebras

The presentation of complex semisimple Lie algebras given by Serre's Theorem can be generalized to construct new families of Lie algebras. Instead of taking the Cartan matrix associated to a root system, one can start with a more general matrix and then use its entries, together with the Serre relations, to define a new Lie algebra. These Lie algebras are usually infinite-dimensional; in fact the finite-dimensional Lie algebras given by this construction are precisely the Lie algebras of types A, B, C, D, E, F, G which we have already seen.

We shall summarize a small section from the introduction of the book *Infinite Dimensional Lie Algebras* by Kac [18]. One defines a *generalised Cartan matrix* to be an $n \times n$ matrix $A = (a_{ij})$ such that

(a) $a_{ij} \in \mathbf{Z}$ for all i, j;

(b) $a_{ii} = 2$, and $a_{ij} \leq 0$ for $i \neq j$;

(c) if $a_{ij} = 0$ then $a_{ji} = 0$.

The associated *Kac–Moody Lie algebra* is the complex Lie algebra over $\mathbf{C}$ generated by the $3n$ elements e_i, f_i, h_i, subject to the Serre relations, as stated in §14.1.2.

When the rank of the matrix A is $n-1$, this construction gives the so-called *affine Kac–Moody Lie algebras*. Modifications of such algebras can be proved to be simple; there is much interest in their representation theory, and several new applications have been discovered.

15.4.1 The Moonshine Conjecture

The largest of the 26 sporadic simple groups is known (because of its enormous size) as the *monster group*. Its order is

$$2^{46}.3^{20}.5^9.7^6.11^2.13^3.17.19.23.29.31.41.47.59.71 \approx 8 \times 10^{53}.$$

Like Lie algebras and associative algebras, groups also have representations. The three smallest representations of the monster group over the complex numbers have dimensions 1, 196883 (it was through this representation that the monster was discovered), and 21296876.

In 1978, John MacKay noticed a near coincidence with the coefficients of the Fourier series expansion of the elliptic modular function j,

$$j(\tau) = q^{-1} + 744 + 196884q + 21493760q^2 + \ldots,$$

where $q = e^{2\pi i\tau}$. As well as noting that $196884 = 196883 + 1$ and $21493760 = 21296876 + 196883 + 1$, he showed that (with a small generalisation) this connection persisted for *all* the coefficients of the j-function.

That there could be an underlying connection between the monster group and modular functions seemed at first so implausible that this became known as the *Moonshine Conjecture*. Yet in 1998 Borcherds succeeded in establishing just such a connection, thus proving the Moonshine Conjecture. A very important part of his work was a further generalisation of the Kac–Moody Lie algebras connected with the exceptional root system of type E_8.

Borcherds was awarded a Fields Medal for his work. A survey can be found in Ray [19]. The reader might also like to read the article by Carter [8].

15.5 The Restricted Burnside Problem

In 1902, William Burnside wrote "A still undecided point in the theory of discontinuous groups is whether the order of a group may not be finite, while the order of every operation it contains is finite." Here we shall consider a variation on his question which can be answered using techniques from Lie algebras.

We must first introduce two definitions: a group G has *exponent* n if $g^n = 1$ for all $g \in G$, and, moreover, n is the least number with this property. A group is *r-generated* if all its elements can be obtained by repeatedly composing a fixed subset of r of its elements. The *restricted Burnside problem* asks: Given $r, n \geq 1$, is there an upper bound on the orders of the finite r-generated groups of exponent n?

Since there are only finitely many isomorphism classes of groups of any given order, the restricted Burnside problem has an affirmative answer if and only if there are (up to isomorphism) only finitely many finite r-generated groups of exponent n.

The reader may well have already seen this problem in the case $n = 2$.

Exercise 15.5

Suppose that G has exponent 2 and is generated by $g_1, \ldots, g_r$. Show that G is abelian and that $|G| \leq 2^r$.

So for $n = 2$, our question has an affirmative answer. In 1992, Zelmanov proved that this is the case whenever n is a prime power. Building on earlier work of Hall and Higman, this was enough to show that the answer is affirmative for all n and r. In 1994, Zelmanov was awarded a Fields Medal for his work.

We shall sketch a proof for the case $n = 3$, which shows some of the ideas in Zelmanov's proof.

Let G be a finitely generated group of exponent p, where p is prime. We define the *lower central series* of G by $G^0 = G$ and $G^i = [G, G^{i-1}]$ for $i \geq 1$. Here $[G, G^{i-1}]$ is the group generated by all group commutators $[x, y] = x^{-1}y^{-1}xy$ for $x \in G, y \in G^{i-1}$. We have

$$G = G^0 \geq G^1 \geq G^2 \geq \ldots.$$

If for some $m \geq 1$ we have $G^m = 1$, then we say G is *nilpotent*.

The notation for group commutators used above is standard; x and y are group elements and the operations are products and inverses in a group. It should not be confused with a commutator in a Lie algebra.

Remark 15.15

It is no accident that the definition of nilpotency for groups mirrors that for Lie algebras. Indeed, nilpotency was first considered for Lie algebras and only much later for groups. This is in contrast to solvability, which was first considered for groups by Galois in his 1830s work on the solution of polynomial equations by radicals.

Each G^i/G^{i+1} is a finitely generated abelian group all of whose non-identity elements have order p. In other words, it is a vector space over $\mathbf{F}_p$, the field with p elements. We may make the (potentially infinite-dimensional) vector space

$$B = \bigoplus_{i=0}^{\infty} G^i/G^{i+1}$$

into a Lie algebra by defining

$$[xG^i, yG^j] = [x, y]G^{i+j}$$

and extending by linearity to arbitrary elements of B. Here on the left we have a commutator in the Lie algebra B and on the right a commutator taken in the group G. It takes some work to see that with this definition the Lie bracket is well defined and satisfies the Jacobi identity — see Vaughan-Lee, *The Restricted Burnside Problem* [24] §2.3, for details. Anticommutativity is more easily seen since if $x, y \in G$, then $[x, y]^{-1} = [y, x]$.

If G is nilpotent (and still finitely generated) then it must be finite, for each G^i/G^{i+1} is a finitely generated abelian group of exponent p, and hence finite. Moreover, if G is nilpotent, then B is a nilpotent Lie algebra. Unfortunately, the converse does not hold because the lower central series might terminate with $G^i = G^{i+1}$ still being an infinite group. However, one can still say something: For the proof of the following theorem, see §2.3 of Vaughan-Lee [24].

Theorem 15.16

If B is nilpotent, then there is an upper bound on the orders of the finite r-generated groups of exponent n. □

The general proof that B is nilpotent is hard. When $p = 3$, however, there are some significant simplifications. By Exercise 4.8, it is sufficient to prove that $[x, [x, y]] = 0$ for all $x, y \in B$. By the construction of B, this will hold if and only if $[g, [g, h]] = 1$ for all $g, h \in G$, now working with group commutators. We now show that this follows from the assumption that G has exponent 3:

$$
\begin{aligned}
[g, [g, h]] &= g^{-1}[g, h]^{-1}g[g, h] \\
&= g^{-1}h^{-1}g^{-1}hggg^{-1}h^{-1}gh \\
&= g^{-1}h^{-1}(g^{-1}hg^{-1})ggh^{-1}gh \\
&= g^{-1}h^{-1}h^{-1}gh^{-1}g^{-1}h^{-1}gh \\
&= g^{-1}hg(h^{-1}g^{-1}h^{-1})gh \\
&= g^{-1}hgghggh \\
&= (g^{-1}hg^{-1})hg^{-1}h \\
&= h^{-1}gh^{-1}hg^{-1}h \\
&= 1,
\end{aligned}
$$

where the bracketing indicates that in the coming step the "rewriting rule" $aba = b^{-1}a^{-1}b^{-1}$ for $a, b \in G$ will be used; this identity holds because $ababab = (ab)^3 = 1$. The reader might like to see if there is a shorter proof.

We must use the elementary argument of Exercise 4.8 rather than Engel's Theorem to prove that B is nilpotent since we have only proved Engel's Theorem for finite-dimensional Lie algebras. In fact, one of Zelmanov's main achievements was to prove an infinite-dimensional version of Engel's Theorem.

15.6 Lie Algebras over Fields of Prime Characteristic

Many Lie algebras over fields of prime characteristic occur naturally; for example, the Lie algebras just seen in the context of the restricted Burnside problem. We have already seen that such Lie algebras have a behaviour different from complex Lie algebras; for example, Lie's Theorem does not hold — see Exercise 6.4. However, other properties appear. For example, let A be an algebra defined over a field of prime characteristic p. Consider the Lie algebra Der A of derivations of A. The Leibniz formula (see Exercise 1.19) tells us that

$$D^p(xy) = \sum_{k=0}^{p} \binom{p}{k} D^k(x) D^{p-k}(y) = x D^p(y) + D^p(x) y$$

for all $x, y \in A$. Thus the p-th power of a derivation is again a derivation. This was one of the examples that led to the formulation of an axiomatic definition of p-*maps* on Lie algebras. A Lie algebra with a p-map is known as a p-Lie algebra. Details of this may be found in Jacobson's book *Lie Algebras* [15] and also in Strade and Farnsteiner, *Modular Lie Algebras and their Representations* [22] or, especially for the representation theory, Jantzen [17].

What can be said about simple Lie algebras over fields of prime characteristic p? Since Lie's Theorem fails in this context, one might expect that the classification of simple Lie algebras over the complex numbers would not generalise. For example, Exercise 15.11 shows that $\mathsf{sl}(n, F)$ is not simple when the characteristic of F divides n. Moreover, new simple Lie algebras have been discovered over fields of prime characteristic that do not have any analogues in characteristic zero.

As an illustration, we shall define the Witt algebra $W(1)$. Fix a field F of characteristic p. The Witt algebra $W(1)$ over F is p-dimensional, with basis

$$e_{-1}, e_0, \ldots, e_{p-2}$$

and Lie bracket

$$[e_i, e_j] = \begin{cases} (j - i) e_{i+j} & -1 \leq i + j \leq p - 2 \\ 0 & \text{otherwise.} \end{cases}$$

When $p = 2$, this algebra is the 2-dimensional non-abelian Lie algebra.

Exercise 15.6

Show that $W(1)$ is simple for $p \geq 3$. Show that if $p = 3$, the Lie algebra $W(1)$ is isomorphic to $\mathsf{sl}(2, F)$. Show that if $p > 3$ then $W(1)$ is not isomorphic to any classical Lie algebra defined over F. *Hint*: The dimensions of the classical Lie algebras (defined over any field) are as given in Exercise 12.1.

A classification of the simple Lie algebras over prime characteristic p is work currently in progress by Premet and Strade.

15.7 Quivers

A *quiver* is another name for a directed graph, for instance,

is a quiver with vertices labelled 1, 2, 3, 4 and arrows labelled α, β, γ. The *underlying graph* of a quiver is obtained by ignoring the direction of the arrows.

A *path* in a quiver is a sequence of arrows which can be composed. In the example above, $\beta\alpha$ is a path (we read paths from right to left as this is the order in which we compose maps), but $\alpha\beta$ and $\alpha\gamma$ are not.

Let $\mathcal{Q}$ be a quiver and let F be a field. The path algebra $F\mathcal{Q}$ is the vector space which has as basis all paths in $\mathcal{Q}$, including the vertices, regarded as paths of length zero. For example, the path algebra of the quiver above has basis

$$\{e_1, e_2, e_3, e_4, \alpha, \beta, \gamma, \beta\alpha, \beta\gamma\}.$$

If two basis elements can be composed to make a path, then their product is defined to be that path. Otherwise, their product is zero. For example, the product of β and α is $\beta\alpha$ since $\beta\alpha$ is a path, whereas the product of α and γ is zero. The behaviour of the vertices is illustrated by $e_1^2 = e_1$, $e_2\alpha = \alpha e_1 = \alpha$, $e_1 e_2 = 0$. This turns $F\mathcal{Q}$ into an associative algebra, which is finite-dimensional precisely when $\mathcal{Q}$ has no oriented cycles.

One would like to understand the representations of $F\mathcal{Q}$. Let V be an $F\mathcal{Q}$-module. The vertices e_i are idempotents whose sum is the identity of the algebra

FQ and $e_i e_j = 0$ if $i \neq j$, so we can use them to decompose V as a direct sum of subspaces

$$V = \bigoplus e_i V.$$

The arrows act as linear maps between the $e_i V$. For example, in the quiver above, $\alpha = e_2 \alpha e_1$ so $\alpha(e_1 V) \subseteq e_2 V$. This allows us to draw a module pictorially: For instance,

$$
\begin{array}{c}
0 \\
\downarrow {\scriptstyle 0} \\
F \xrightarrow{\ 1\ } F \xrightarrow{\ 0\ } 0
\end{array}
$$

shows a 2-dimensional module V, where $e_1 V \cong e_2 V \cong F$ and α acts as an isomorphism between $e_1 V$ and $e_2 V$ (and β and γ act as the zero map).

The simple FQ-modules are all 1-dimensional, with one for each vertex. For example,

$$
\begin{array}{c}
0 \\
\downarrow {\scriptstyle 0} \\
0 \xrightarrow{\ 0\ } F \xrightarrow{\ 0\ } 0
\end{array}
$$

shows the simple module corresponding to vertex 2. In this module, $e_2 V = V$ and all the other basis elements act as 0.

Usually there will be FQ-modules which are not direct sums of simple modules. For example, the first module defined above has $e_2 V$ as its unique non-trivial submodule, and so it does not split up as a direct sum of simple modules. Thus there are indecomposable FQ-modules which are not irreducible. One can measure the extent to which complete reducibility fails to hold by asking how many indecomposable FQ-modules there are.

If there are only finitely indecomposable modules (up to isomorphism), the algebra FQ is said to have *finite type*. In the 1970s, Gabriel found a necessary and sufficient condition for a quiver algebra to have finite type. He proved the following theorem.

Theorem 15.17 (Gabriel's Theorem)

The path algebra FQ has finite type if and only if the underlying graph of Q is a disjoint union of Dynkin diagrams of types A, D, E. Moreover, the indecomposable KQ-modules are parametrized naturally by the positive roots of the associated root system.

Example 15.18

Consider the quiver of type A_4

$$\underset{1}{\cdot} \xrightarrow{\alpha_1} \underset{2}{\cdot} \xrightarrow{\alpha_2} \underset{3}{\cdot} \xrightarrow{\alpha_3} \underset{4}{\cdot}$$

By Gabriel's Theorem, the indecomposable representations of this quiver are in bijection with the positive roots in the root system of type A_4. The simple roots $\alpha_1, \alpha_2, \alpha_3, \alpha_4$ correspond to the simple modules. The positive root $\alpha_1 + \alpha_2$ corresponds to the module

$$F \xrightarrow{\ 1\ } F \xrightarrow{\ 0\ } 0 \xrightarrow{\ 0\ } 0$$

and so on.

One might wonder whether this connection with Dynkin diagrams is merely an accident. Not long ago, Ringel discovered a deep connection between quivers and the theory of Lie algebras. He showed that, when F is a finite field, one may define an algebra which encapsulates all the representations of FQ. This algebra is now known as the *Ringel–Hall algebra*; Ringel proved that this algebra is closely related to the quantum group of the same type as the underlying graph of the quiver.

EXERCISES

15.7.   Tensor products can also be used to construct representations of a direct sum of two Lie algebras. Let L_1 and L_2 be isomorphic copies of $\mathsf{sl}(2, \mathbf{C})$ and let $L = L_1 \oplus L_2$. Let $V(a)$ and $V(b)$ be irreducible modules for $\mathsf{sl}(2, \mathbf{C})$ with highest weights a and b, respectively.

(i) Show that we may make $V(a) \otimes V(b)$ into a module for L by setting

$$(x, y) \cdot v \otimes w = ((x \cdot v) \otimes w) + (v \otimes (y \cdot w))$$

for $x \in L_1$, $y \in L_2$, $v \in V(a)$, and $w \in V(b)$.

(ii) Show that $V(a) \otimes V(b)$ is an irreducible representation of L with highest weight λ defined by

$$\lambda(h, 0) = a \quad \lambda(0, h) = b.$$

It can be shown that this construction gives every irreducible L-module. By Exercise 10.8, $\mathsf{sl}(2,\mathbf{C}) \oplus \mathsf{sl}(2,\mathbf{C}) \cong \mathsf{so}(4,\mathbf{C})$, so we have constructed all the finite-dimensional representations of $\mathsf{so}(4,\mathbf{C})$. Generalising these ideas, one can show that any semisimple Lie algebra has a faithful irreducible representation; from this it is not hard to prove a (partial) converse of Exercise 12.4.

15.8.   Let L be a Lie algebra and let $U(L)$ be its universal enveloping algebra as defined above. Let $\iota : L \to U(L)$ be the linear map defined by $\iota(x_i) = X_i$.

Let A be an associative algebra; we may also view A as a Lie algebra with Lie bracket $[x,y] = xy - yx$ for $x,y \in A$ (see §1.5).

(i) Show that $U(L)$ has the following *universal property*: Given a Lie algebra homomorphism $\varphi : L \to A$, there exists a *unique* homomorphism of associative algebras $\theta : U(L) \to A$ such that $\theta \circ \iota = \varphi$. In other words, the following diagram commutes:

(ii) Suppose that V is an associative algebra and $\iota' : L \to V$ is a Lie algebra homomorphism (where we regard V as a Lie algebra) such that if we replace ι with ι' and $U(L)$ with V in the commutative diagram above then V has the universal property of $U(L)$. Show that V and $U(L)$ are isomorphic. In particular, this shows that $U(L)$ does not depend on the choice of basis of L.

(iii) Let $x_1, \ldots, x_k \in L$. Suppose that L has a representation $\varphi : L \to \mathsf{gl}(V)$ such that $\varphi(x_1), \ldots, \varphi(x_k)$ are linearly independent. Show that $X_1, \ldots, X_k$ are linearly independent elements of $U(L)$. Hence prove that if L is semisimple then ι is injective.

15.9.   Let $\delta : L \to L$ be a derivation of a complex finite-dimensional Lie algebra L. Suppose that $\delta^n = 0$ where $n \geq 1$. Define $\exp(\delta) : L \to L$ by

$$\exp(\delta)(x) = \left(1 + \delta + \frac{\delta^2}{2!} + \ldots\right)x.$$

(By hypothesis the sum is finite.) Prove that $\exp(\delta)$ is an *automorphism* of L; that is, $\exp(\delta) : L \to L$ is an invertible linear map such that

$$[\exp \delta(x), \exp \delta(y)] = \exp \delta([x,y]) \quad \text{for all } x, y \in L.$$

15.10.   Let L be a finite-dimensional complex Lie algebra and let α be an automorphism of L. For $\nu \in \mathbf{C}$, let

$$L_\nu = \{x \in L : \alpha(x) = \nu(x)\}.$$

Show that $[L_\lambda, L_\mu] \subseteq L_{\lambda\mu}$. Now suppose that we have $\alpha^3 = 1$, and that α fixes no non-zero element of L. Prove that L is nilpotent.

15.11.   Let F be a field of prime characteristic p. Show that if p divides n then $\mathsf{sl}(n, F)$ is not simple.

16

Appendix A: Linear Algebra

This appendix gives a summary of the results we need from linear algebra. Recommended for further reading are Blyth and Robertson's books *Basic Linear Algebra* [4] and *Further Linear Algebra* [5] and Halmos *Finite-Dimensional Vector Spaces* [11].

We expect that the reader will already know the definition of vector spaces and will have seen some examples. For most of this book, we deal with finite-dimensional vector spaces over the complex numbers, so the main example to bear in mind is $\mathbf{C}^n$, which we think of as a set of column vectors.

We assume that the reader knows about bases, subspaces, and direct sums. We therefore begin our account by describing quotient spaces. Next we discuss the connection between linear maps and matrices, diagonalisation of matrices, and Jordan canonical form. We conclude by reviewing the bilinear algebra needed in the main text.

16.1 Quotient Spaces

Suppose that W is a subspace of the vector space V. A *coset of W* is a set of the form

$$v + W := \{v + w : w \in W\}.$$

It is important to realise that unless $W = 0$, each coset will have many different labels; in fact, $v + W = v' + W$ if and only if $v - v' \in W$.

The *quotient space* V/W is the set of all cosets of W. This becomes a vector space, with zero element $0 + W = W$, if addition is defined by

$$(v + W) + (v' + W) := (v + v') + W \quad \text{for } v, v' \in V$$

and scalar multiplication by

$$\lambda(v + W) := \lambda v + W \quad \text{for } v, v' \in V, \lambda \in F.$$

One must check that these operations are *well-defined*; that is, they do not depend on the choice of labelling elements. Suppose for instance that $v + W = v' + W$. Then, since $v - v' \in W$, we have $\lambda v - \lambda v' \in W$ for any scalar λ, so $\lambda v + W = \lambda v' + W$.

The following diagram shows the elements of $\mathbf{R}^2/W$, where W is the subspace of $\mathbf{R}^2$ spanned by $\binom{1}{1}$.

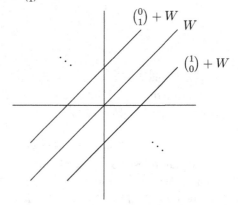

The cosets $\mathbf{R}^2/W$ are all the translations of the line W. One can choose a standard set of coset representatives by picking any line through 0 (other than W) and looking at its intersection points with the cosets of W; this gives a geometric interpretation of the isomorphism $\mathbf{R}^2/W \cong \mathbf{R}$.

It is often useful to consider quotient spaces when attempting a proof by induction on the dimension of a vector space. In this context, it can be useful to know that if $v_1, \ldots, v_k$ are vectors in V such that the cosets $v_1 + W, \ldots, v_k + W$ form a basis for the quotient space V/W, then $v_1, \ldots, v_k$, together with any basis for W, forms a basis for V.

16.2 Linear Maps

Let V and W be vector spaces over a field F. A *linear map* (or *linear transformation*) $x : V \to W$ is a map satisfying

$$x(\lambda u + \mu v) = \lambda x(u) + \mu x(v) \quad \text{for all } u, v \in V \text{ and } \lambda, \mu \in F.$$

A bijective linear map between two vector spaces is an *isomorphism*. We assume the reader knows about the definitions of the image and kernel of a linear map, and can prove the rank-nullity theorem,

$$\dim V = \dim \operatorname{im} x + \dim \ker x.$$

A corollary of the rank-nullity theorem is that if $\dim V = \dim W$ and $x : V \to W$ is injective, then, since $\dim \operatorname{im} x = \dim V$, x is an isomorphism. One can draw the same conclusion if instead we know that x is surjective. We refer to this type of reasoning as an argument by *dimension counting*.

We can now state the isomorphism theorems for vector spaces.

Theorem 16.1 (Isomorphism theorems for vector spaces)

(a) If $x : V \to W$ is a linear map, then $\ker x$ is a subspace of V, $\operatorname{im} x$ is a subspace of W, and

$$V/\ker x \cong \operatorname{im} x.$$

(b) If U and W are subspaces of a vector space, $(U + W)/W \cong U/(U \cap W)$.

(c) Suppose that U and W are subspaces of a vector space V such that $U \subseteq W$. Then W/U is a subspace of V/U and $(V/U)/(W/U) \cong V/W$.

Proof

For part (a), define a map $\varphi : V/\ker x \to \operatorname{im} x$ by

$$\varphi(v + \ker x) = x(v).$$

This map is well-defined since if $v + \ker x = v' + \ker x$ then $v - v' \in \ker x$, so $\varphi(v + \ker x) = x(v) = x(v') = \varphi(v' + \ker x)$. It is routine to check that φ is linear, injective, and surjective, so it gives the required isomorphism.

To prove (b), consider the composite of the inclusion map $U \to U + W$ with the quotient map $U + W \to (U + W)/W$. This gives us a linear map $U \to (U + W)/W$. Under this map, $x \in U$ is sent to $0 \in (U + W)/W$ if and only if $x \in W$, so its kernel is $U \cap W$. Now apply part (a).

Part (c) can be proved similarly; we leave this to the reader. $\qquad\square$

Parts (a), (b) and (c) of this theorem are known respectively as the *first*, *second*, and *third isomorphism theorems*. See Exercise 16.5 for one application.

16.3 Matrices and Diagonalisation

Suppose that $x : V \to V$ is a linear transformation of a finite-dimensional vector space V. Let $\{v_1, \ldots, v_n\}$ be a basis of V. Using this basis, we may define scalars a_{ij} by

$$x(v_j) = \sum_{i=1}^{n} a_{ij} v_i.$$

We say that the $n \times n$ matrix A with entries (a_{ij}) is the *matrix of x* with respect to our chosen basis. Conversely, given a basis of V and a matrix A, we can use the previous equation to define a linear map x, whose matrix with respect to this basis is A.

Exercise 16.1

(i) Let $x : V \to V$ and $y : V \to V$ be linear maps with matrices A and B with respect to a basis of V. Show that, with respect to this basis, the matrix of the composite map yx is the matrix product BA.

(ii) Suppose that x has matrix A with respect to the basis $v_1, \ldots, v_n$ of V. Let $w_1, \ldots, w_n$ be another basis of V. Show that the matrix of A in this new basis is $P^{-1}AP$ where the matrix $P = (p_{ij})$ is defined by

$$w_j = \sum_{i=1}^{n} p_{ij} v_i.$$

Matrices related in this way are said to be *similar*.

It had been said that "a true gentleman never takes bases unless he really has to." We generally agree with this sentiment, preferring to use matrices only when they are necessary for explicit computations (for example in Chapter 12 when we look at the classical Lie algebras). When we are obliged to consider matrices, then we can at least try to choose bases so that they are of a convenient form.

Recall that a non-zero vector $v \in V$ such that $x(v) = \lambda v$ is said to be an *eigenvector* of x with corresponding *eigenvalue* λ. The *eigenspace* for eigenvalue λ is the vector subspace

$$\{v \in V : x(v) = \lambda v\}.$$

It is an elementary fact that non-zero vectors in different eigenspaces are linearly independent. (This will often be useful for us; for example, see step 1 in the proof of Theorem 8.5.)

The linear map x can be represented by a diagonal matrix if and only if V has a basis consisting of eigenvectors for x. This is the same as saying that the space V is a direct sum of x-eigenspaces,

$$V = V_{\lambda_1} \oplus V_{\lambda_2} \oplus \ldots \oplus V_{\lambda_r},$$

where the λ_i are the distinct eigenvalues of x. If this is the case, we say that x is *diagonalisable*.

Note that $\lambda \in \mathbf{C}$ is an eigenvalue of x if and only if $\ker(x - \lambda 1_V)$ is non-zero, which is the case if and only if $\det(x - \lambda 1_V) = 0$. The eigenvalues of x are therefore the roots of the *characteristic polynomial* of x, defined by

$$c_x(X) = \det(x - X 1_V),$$

where X is an indeterminant. Since over $\mathbf{C}$ any non-constant polynomial has a root, this shows that any linear transformation of a complex vector space has an eigenvalue.

The characteristic polynomial of x does not in itself give enough information to determine whether x is diagonalisable — consider for example the matrices

$$\begin{pmatrix} 1 & 0 \\ 0 & 1 \end{pmatrix}, \quad \begin{pmatrix} 1 & 1 \\ 0 & 1 \end{pmatrix}.$$

To get further, one needs the minimal polynomial. The *minimal polynomial* of x is the monic polynomial of least degree which kills x, so $m(X) = X^d + a_{d-1}X^{d-1} + \ldots + a_1 X + a_0$ is the minimal polynomial of x if

$$x^d + a_{d-1}x^{d-1} + \ldots + a_1 x + a_0 1_V = 0$$

and the degree d is as small as possible.

An important property of the minimal polynomial is that if $f(X)$ is any polynomial such that $f(x) = 0$ then $m(X)$ divides $f(X)$.

Exercise 16.2

Prove this assertion by using polynomial division to write $f(X) = a(X)m(X) + r(X)$, where the remainder polynomial $r(X)$ is either 0 or has degree less than that of $m(X)$, and then showing that $r(x) = 0$.

By the famous theorem of Cayley–Hamilton (see Exercise 16.4), the minimal polynomial of x divides the characteristic polynomial of x. We now explore some of the arguments in which the minimal polynomial is used.

16.3.1 The Primary Decomposition Theorem

Theorem 16.2 (Primary decomposition theorem)

Suppose the minimal polynomial of x factorises as

$$(X - \lambda_1)^{a_1} \ldots (X - \lambda_r)^{a_r},$$

where the λ_i are distinct and each $a_i \geq 1$. Then V decomposes as a direct sum of x-invariant subspaces V_i,

$$V = V_1 \oplus V_2 \oplus \ldots \oplus V_r,$$

where $V_i = \ker(x - \lambda_i 1_V)^{a_i}$. The subspaces V_i are said to be the *generalised eigenspaces* of x.

This theorem may be proved by repeatedly applying the following lemma.

Lemma 16.3

If $f(X) \in \mathbb{C}[X]$ and $g(X) \in \mathbb{C}[X]$ are coprime polynomials such that $f(x)g(x) = 0$, then im $f(x)$ and im $g(x)$ are x-invariant subspaces of V. Moreover,

(i) $V = \operatorname{im} f(x) \oplus \operatorname{im} g(x)$, and

(ii) $\operatorname{im} f(x) = \ker g(x)$ and $\operatorname{im} g(x) = \ker f(x)$.

Proof

If $v = f(x)w$, then $xv = f(x)xw$, so the subspaces im $f(x)$ and im $g(x)$ are x-invariant. By Euclid's algorithm, there exist polynomials $a(X), b(X) \in \mathbf{C}[X]$ such that $a(X)f(X) + b(X)g(X) = 1$, so for any $v \in V$,

$$f(x)(a(x)v) + g(x)(b(x)v) = v. \qquad (\star)$$

This shows that $V = \operatorname{im} f(x) + \operatorname{im} g(x)$. If $v \in \operatorname{im} g(x)$ with, say, $v = g(x)w$, then $f(x)v = f(x)g(x)w = 0$, so $\operatorname{im} g(x) \subseteq \ker f(x)$. On the other hand, if $f(x)v = 0$, then by $(\star)$, $v = g(x)(b(x)v)$ so $v \in \operatorname{im} g(x)$. Finally, if

$$v \in \operatorname{im} f(x) \cap \operatorname{im} g(x) = \ker f(x) \cap \ker g(x),$$

then as $f(x)a(x)v = a(x)f(x)v = 0$ and similarly $b(x)g(x) = 0$, it follows from $(\star)$ that $v = 0$. □

The following criterion for a linear map to be diagonalisable follows directly from the primary decomposition theorem.

Theorem 16.4

Let $x : V \to V$ be a linear map of a vector space V. Then x is diagonalisable if and only if the minimal polynomial of x splits as a product of distinct linear factors. $\square$

Corollary 16.5

Let $x : V \to V$ be a diagonalisable linear transformation. Suppose that U is a subspace of V which is invariant under x, that is, $x(u) \in U$ for all $u \in U$.

(a) The restriction of x to U is diagonalisable.

(b) Given any basis of U consisting of eigenvectors for x, we may extend this basis to a basis of V consisting of eigenvectors for x.

Proof

Let $m(X)$ be the minimal polynomial of $x : V \to V$. Let $m_U(X)$ be the minimal polynomial of x, regarded just as a linear transformation of U. Then $m(x)(U) = 0$, so $m_U(X)$ must divide $m(X)$. Hence $m_U(X)$ is a product of distinct linear factors.

Now let $V = V_{\lambda_1} \oplus \ldots \oplus V_{\lambda_r}$ be the decomposition of V into distinct eigenspaces of x. Since x acts diagonalisably on U we have

$$U = U \cap V_{\lambda_1} \oplus \ldots \oplus U \cap V_{\lambda_r}.$$

Extend the basis of each $U \cap V_{\lambda_i}$ to a basis of V_{λ_i}. This gives us a basis of V of the required form. $\square$

We now give another application of the primary decomposition theorem.

Lemma 16.6

Suppose that x has minimal polynomial

$$f(X) = (X - \lambda_1)^{a_1} \ldots (X - \lambda_r)^{a_r},$$

where the λ_i are pairwise distinct. Let the corresponding primary decomposition of V as a direct sum of generalised eigenspaces be

$$V = V_1 \oplus \ldots \oplus V_r,$$

where $V_i = \ker(x - \lambda_i 1_V)^{a_i}$. Then, given any $\mu_1, \ldots, \mu_r \in \mathbb{C}$, there is a polynomial $p(X)$ such that

$$p(x) = \mu_1 1_{V_1} + \mu_2 1_{V_2} \ldots + \mu_r 1_{V_r}.$$

Proof

Suppose we could find a polynomial $f(X) \in \mathbf{C}[X]$ such that

$$f(X) \equiv \mu_i \bmod (X - \lambda_i)^{a_i}.$$

Take $v \in V_i = \ker(x - \lambda_i 1_V)^{a_i}$. By our supposition, $f(X) = \mu_i + a(X)(X - \lambda_i)^{a_i}$ for some polynomial $a(X)$. Hence

$$f(x)v = \mu_i 1_{V_i} v + a(x)(x - \lambda_i)^{a_i} v = \mu_i v,$$

as required.

The polynomials $(X - \lambda_1)^{a_1} \ldots, (X - \lambda_r)^{a_r}$ are coprime. We may therefore apply the Chinese Remainder Theorem, which states that in these circumstances the map

$$\mathbf{C}[X] \to \bigoplus_{i=1}^{r} \frac{\mathbf{C}[X]}{(X - \lambda_i)^{a_i}}$$

$$f(X) \mapsto (f(X) \bmod (X - \lambda_1)^{a_1}, \ldots, f(X) \bmod (X - \lambda_r)^{a_r})$$

is surjective, to obtain a suitable $p(X)$. $\square$

In terms of matrices, this lemma says that

$$p(x) = \begin{pmatrix} \mu_1 I_{n_1} & 0 & \ldots & 0 \\ 0 & \mu_2 I_{n_2} & \ldots & 0 \\ \vdots & \vdots & \ddots & \vdots \\ 0 & 0 & \ldots & \mu_r I_{n_r} \end{pmatrix},$$

where $n_i = \dim V_i$ and I_s denotes the $s \times s$ identity matrix.

16.3.2 Simultaneous Diagonalisation

In the main text, we shall several times have a finite family of linear transformations of a vector space V, each of which is individually diagonalisable. When can one find a basis of V in which they are all simultaneously diagonal?

Lemma 16.7

Let $x_1, \ldots, x_k : V \to V$ be diagonalisable linear transformations. There is a basis of V consisting of simultaneous eigenvectors for all the x_i if and only if they commute. (That is, $x_i x_j = x_j x_i$ for all pairs i, j.)

Proof

For the "only if" direction we note that diagonal matrices commute with one another, so if we can represent all the x_i by diagonal matrices, they must commute.

The main step in the "if" direction is the case $k = 2$. Write V as a direct sum of eigenspaces for x_1, say $V = V_{\lambda_1} \oplus \ldots \oplus V_{\lambda_r}$, where the λ_i are the distinct eigenvalues of x_1. If $v \in V_{\lambda_i}$ then so is $x_2(v)$, for

$$x_1 x_2(v) = x_2 x_1(v) = x_2(\lambda_i v) = \lambda_i(x_2(v)).$$

We now apply Corollary 16.5(a) to deduce that x_2 restricted to V_{λ_i} is diagonalisable. A basis of V_{λ_i} consisting of eigenvectors for x_2 is automatically a basis of eigenvectors for x_1, so if we take the union of a basis of eigenvectors for x_2 on each V_{λ_i}, we get a basis of V consisting of simultaneous eigenvectors for both x_1 and x_2.

The inductive step is left to the reader. $\square$

In Exercise 16.6, we give a small generalisation which will be needed in the main text.

16.4 Interlude: The Diagonal Fallacy

Consider the following (fallacious) argument. Let V be a 2-dimensional vector space, say with basis v_1, v_2. Let $x : V \to V$ be the linear map whose matrix with respect to this basis is

$$\begin{pmatrix} 0 & 1 \\ 0 & 0 \end{pmatrix}.$$

We claim that if U is a subspace of V such that $x(U) \subseteq U$, then either $U = 0$, $U = \mathrm{Span}\{v_1\}$, or $U = V$. Clearly each of these subspaces is invariant under x, so we only need to prove that there are no others. But since $x(v_2) = v_1$, $\mathrm{Span}\{v_2\}$ is not x-invariant. (QED?)

Here we committed the *diagonal fallacy*: We assumed that an arbitrary subspace of V would contain one of our chosen basis vectors. This assumption is very tempting — which perhaps explains why it is so often made — but it is nonetheless totally unjustified.

Exercise 16.3

Give a correct proof of the previous result.

The following diagram (which is frequently useful as a counterexample in linear algebra) illustrates how the fallacy we have been discussing gets its name.

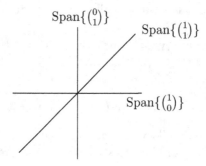

16.5 Jordan Canonical Form

Let V be a finite-dimensional complex vector space and let $x : V \to V$ be a linear map. Exercise 6.2 outlines the proof that one can always find a basis of V in which x is represented by an upper triangular matrix. For many purposes, this result is sufficient. For example, it implies that a nilpotent map may be represented by a strictly upper triangular matrix, and so nilpotent maps have trace 0.

Sometimes, however, one needs the full strength of Jordan canonical form. A general matrix in Jordan canonical form looks like

$$\begin{pmatrix} A_1 & 0 & \cdots & 0 \\ 0 & A_2 & \cdots & 0 \\ \vdots & \vdots & \ddots & \vdots \\ 0 & 0 & \cdots & A_r \end{pmatrix},$$

where each A_i is a *Jordan block matrix* $J_t(\lambda)$ for some $t \in \mathbb{N}$ and $\lambda \in \mathbb{C}$:

$$J_t(\lambda) = \begin{pmatrix} \lambda & 1 & 0 & \cdots & 0 & 0 \\ 0 & \lambda & 1 & \cdots & 0 & 0 \\ 0 & 0 & \lambda & \cdots & 0 & 0 \\ \vdots & \vdots & \vdots & \ddots & \vdots & \vdots \\ 0 & 0 & 0 & \cdots & \lambda & 1 \\ 0 & 0 & 0 & \cdots & 0 & \lambda \end{pmatrix}_{t \times t}.$$

We now outline a proof that any linear transformation of a complex vector space can be represented by a matrix in Jordan canonical form.

The first step is to reduce to the case where $x^q = 0$ for some $q \geq 1$; that is, x is a nilpotent linear map.

By the primary decomposition theorem, it suffices to consider the case where x has only one eigenvalue, say λ. Then by considering $x - \lambda 1_V$, we may reduce to the case where x acts nilpotently. So it suffices to show that a nilpotent transformation can be put into Jordan canonical form.

16.5.1 Jordan Canonical Form for Nilpotent Maps

We shall work by induction on $\dim V$.

Suppose that $x^q = 0$ and $x^{q-1} \neq 0$. Let $v \in V$ be any vector such that $x^{q-1}v \neq 0$. One can check that the vectors $v, xv, \ldots, x^{q-1}v$ are linearly independent. Their span, U say, is an x-invariant subspace of V. With respect to the given basis of U, the matrix of $x : U \to U$ is the $q \times q$ matrix

$$J_q(0) = \begin{pmatrix} 0 & 1 & 0 & \ldots & 0 \\ 0 & 0 & 1 & \ldots & 0 \\ \vdots & \vdots & \vdots & \ddots & \vdots \\ 0 & 0 & 0 & \ldots & 1 \\ 0 & 0 & 0 & \ldots & 0 \end{pmatrix}.$$

Suppose we can find an x-invariant complementary subspace to U; that is, a subspace C such that x maps C into C and $V = U \oplus C$. Then, by induction, there is a basis of C in which the matrix of x restricted to C is in Jordan canonical form. Putting the bases of C and U together gives us a suitable basis for V.

To show that a suitable complement exists, we use a further induction on q. If $q = 1$, then $x = 0$ and any vector space complement to $\mathrm{Span}\{v\}$ will do. Now suppose we can find complements when $x^{q-1} = 0$.

Consider $\mathrm{im}\, x \subseteq V$. On $\mathrm{im}\, x$, x acts as a nilpotent linear map whose $q - 1$ power is 0, so by induction on q we get

$$\mathrm{im}\, x = \mathrm{Span}\left\{xv, \ldots, x^{q-1}v\right\} \oplus W$$

for some x-invariant subspace W. Note that $U \cap W = 0$. Our task is to extend W to a suitable x-invariant complement for U in V.

Suppose first that $W = 0$. In this case, $\mathrm{im}\, x = \mathrm{Span}\left\{xv, \ldots, x^{q-1}v\right\}$ and $\ker x \cap \mathrm{im}\, x = \left\langle x^{q-1}v \right\rangle$. Extend $x^{q-1}v$ to a basis of $\ker x$, say by $v_1, \ldots, v_s$. By the rank-nullity formula

$$v, xv \ldots, x^{q-1}v, v_1, \ldots, v_s$$

is a basis of V. The subspace spanned by $v_1, \ldots, v_s$ is an x-invariant complement to U.

Now suppose that $W \neq 0$. Then x induces a linear transformation, say $\bar{x}$, on V/W. Let $\bar{v} = v + W$. Since $\operatorname{im} \bar{x} = \operatorname{Span} \{\bar{x}\bar{v}, \ldots, \bar{x}^{q-1}\bar{v}\}$, the first case implies that there is an $\bar{x}$-invariant complement in V/W to $\operatorname{Span} \{\bar{v}, \bar{x}\bar{v}, \ldots \bar{x}^{q-1}\bar{v}\}$. The preimage of this complement in V is a suitable complement to U.

16.6 Jordan Decomposition

Any linear transformation x of a complex vector space V has a *Jordan decomposition*, $x = d + n$, where d is diagonalisable, n is nilpotent, and d and n commute.

One can see this by putting x into Jordan canonical form: Fix a basis of V in which x is represented by a matrix in Jordan canonical form. Let d be the map whose matrix in this basis has the diagonal entries of x down its diagonal, and let $n = x - d$. For example we might have

$$x = \begin{pmatrix} 1 & 1 & 0 \\ 0 & 1 & 0 \\ 0 & 0 & 1 \end{pmatrix}, \quad d = \begin{pmatrix} 1 & 0 & 0 \\ 0 & 1 & 0 \\ 0 & 0 & 1 \end{pmatrix}, \quad n = \begin{pmatrix} 0 & 1 & 0 \\ 0 & 0 & 0 \\ 0 & 0 & 0 \end{pmatrix}.$$

As n is represented by a strictly upper triangular matrix, it is nilpotent. We leave it to the reader to check that d and n commute.

In applications it is useful to know that d and n can be expressed as polynomials in x. In the following lemma, we also prove a related result that is needed in Chapter 9.

Lemma 16.8

Let x have Jordan decomposition $x = d + n$ as above, where d is diagonalisable, n is nilpotent, and d, n commute.

(a) There is a polynomial $p(X) \in \mathbf{C}[X]$ such that $p(x) = d$.

(b) Fix a basis of V in which d is diagonal. Let $\bar{d}$ be the linear map whose matrix with respect to this basis is the complex conjugate of the matrix of d. There is a polynomial $q(X) \in \mathbf{C}[X]$ such that $q(x) = \bar{d}$.

Proof

Let $\lambda_1, \ldots, \lambda_r$ be the distinct eigenvalues of x. The minimal polynomial of x is then

$$m(X) = (X - \lambda_1)^{a_1} \ldots (X - \lambda_r)^{a_r},$$

where a_i is the size of the largest Jordan block with eigenvalue λ_i.

We can now apply Lemma 16.6 to get the polynomials we seek. For part (a) take $\mu_i = \lambda_i$, and for part (b) take $\mu_i = \bar{\lambda}_i$.   □

Part (a) of this lemma can be used to prove that the Jordan decomposition of a linear map is unique — see Exercise 16.7 below.

16.7 Bilinear Algebra

As well as the books already mentioned, we recommend Artin's *Geometric Algebra* [1] for further reading on bilinear algebra. From now on, we let V be an n-dimensional vector space over a field F.

16.7.1 Dual spaces

The *dual space* of V, denoted $V^\star$, is by definition the set of all linear maps from V to F. Thus, if $f, g \in V^\star$, then $f + g$ is defined by $(f + g)(v) = f(v) + g(v)$ for $v \in V$, and if $\lambda \in F$, then λf is defined by $(\lambda f)(v) = \lambda f(v)$.

Given a vector space basis $\{v_1, \ldots, v_n\}$ of V, one defines the associated *dual basis* of V as follows. Let $f_i : V \to F$ be the linear map defined on basis elements by

$$f_i(v_j) = \begin{cases} 1 & i = j \\ 0 & i \neq j. \end{cases}$$

It is not hard to check that $f_1, \ldots, f_n$ is a basis for $V^\star$. In particular $\dim V = \dim V^\star$.

The dual space of V^* can be identified with V in a natural way. Given $v \in V$, we may define an *evaluation map* $\varepsilon_v : V^* \to F$ by

$$\varepsilon_v(f) := f(v) \quad \text{for all } f \in V^*.$$

It is straightforward to check that ε_v is linear and so belongs to the dual space of $V^\star$; that is, to $V^{\star\star}$. Moreover, the map $v \mapsto \varepsilon_v$ (which we might call ε) from V to $V^{\star\star}$ is itself linear. We claim that $\varepsilon : V \to V^{\star\star}$ is an isomorphism.

Since we have already shown that $\dim V = \dim V^* = \dim V^{**}$, it is sufficient to show that $\varepsilon_v = 0$ implies $v = 0$. One way to do this is as follows. If $v \neq 0$, then we may extend v to a basis of V and take the associated dual basis. Then $f_1(v) = 1$ and hence $\varepsilon_v(f_1) \neq 0$, so $\varepsilon_v \neq 0$.

If U is a subspace of V we let

$$U^\circ = \{f \in V^* : f(u) = 0 \text{ for all } u \in U\}$$

be the *annihilator* of U in V^*. One can show that U° is a subspace of V^* and that

$$\dim U + \dim U^\circ = \dim V.$$

A proof of the last statement is outlined in Exercise 16.8.

Given a subspace W of V^*, we can similarly define the annihilator of W in V^{**}. Under the identification of V^{**} with V, the annihilator of W becomes

$$W^0 = \{v \in V : f(v) = 0 \text{ for all } f \in W\}.$$

In particular, we have $\dim W + \dim W^0 = \dim V$.

16.7.2 Bilinear Forms

Definition 16.9

A *bilinear form* on V is a map

$$(-, -) : V \times V \to F$$

such that

$$(\lambda_1 v_1 + \lambda_2 v_2, w) = \lambda_1(v_1, w) + \lambda_2(v_2, w),$$
$$(v, \mu_1 w_1 + \mu_2 w_2) = \mu_1(v, w_1) + \mu_2(v, w_2),$$

for all $v, w, v_i, w_i \in V$ and $\lambda_i, \mu_i \in F$.

For example, if $F = \mathbf{R}$ and $V = \mathbf{R}^n$, then the usual dot product is a bilinear form on V.

As for linear transformations, we can represent bilinear forms by matrices. Suppose that $(-, -)$ is a bilinear form on the vector space V and that V has basis $\{v_1, \ldots, v_n\}$. The *matrix of* $(-, -)$ with respect to this basis is $A = (a_{ij})$, where $a_{ij} = (v_i, v_j)$. If we change the basis, say to $\{w_1, \ldots, w_n\}$, then the new matrix representing $(-, -)$ is $P^t A P$ where $P = (p_{ij})$ is the $n \times n$ matrix defined by

$$w_j = \sum_{i=1}^{n} p_{ij} v_i.$$

Matrices related in this way are said to be *congruent*.

Conversely, given an $n \times n$ matrix $S = (s_{ij})$, we may define a bilinear form on V by setting

$$(v_i, v_j) = s_{ij}$$

and extending "bilinearly" to arbitrary elements in $V \times V$. That is, if $v = \sum_i \lambda_i v_i$ and $w = \sum_j \mu_j v_j$ with λ_i and μ_j scalars, then

$$(v, w) = \sum_{i=1}^{n} \sum_{j=1}^{n} s_{ij} \lambda_i \mu_j.$$

The last equation may be written in matrix form as

$$(v, w) = (\lambda_1 \ldots \lambda_n) \begin{pmatrix} s_{11} & \cdots & s_{1n} \\ \vdots & \ddots & \vdots \\ s_{n1} & \cdots & s_{nn} \end{pmatrix} \begin{pmatrix} \mu_1 \\ \vdots \\ \mu_n \end{pmatrix}.$$

Given a subset U of V, we set

$$U^{\perp} := \{v \in V : (u, v) = 0 \text{ for all } u \in U\}.$$

This is always a subspace of V. We say that the form $(-, -)$ is *non-degenerate* if $V^{\perp} = \{0\}$.

Example 16.10

Let U be a $2m$-dimensional vector space with basis $u_1, \ldots, u_{2m}$, and let

$$S = \begin{pmatrix} 0 & I_m \\ I_m & 0 \end{pmatrix},$$

where I_m is the identity matrix of size $m \times m$. The bilinear form associated to S may be shown to be non-degenerate. (For example, this follows from Exercise 16.9.) However, the restriction of the form to the subspace spanned by $u_1, \ldots, u_m$ is identically zero.

For a more substantial example, see Exercise 16.10 below.

We now explain the connection between bilinear forms and dual spaces. Let $\varphi : V \to V^{\star}$ be the linear map defined by $\varphi(v) = (-, v)$. That is, $\varphi(v)$ is the linear map sending $u \in V$ to (u, v). If $(-, -)$ is non-degenerate, then $\ker \varphi = 0$, so by dimension counting, φ is an isomorphism. Thus every element of $V^{\star}$ is of the form $(-, v)$ for a unique $v \in V$; this is a special case of the *Riesz representation theorem*. A small generalisation of this argument can be used to prove the following lemma — see Exercise 16.8.

Lemma 16.11

Suppose that $(-,-)$ is a non-degenerate bilinear form on the vector space V. Then, for all subspaces U of V, we have

$$\dim U + \dim U^\perp = \dim V.$$

If $U \cap U^\perp = 0$, then $V = U \oplus U^\perp$ and, furthermore, the restrictions of $(-,-)$ to U and to $U^\perp$ are non-degenerate. $\qquad\square$

16.7.3 Canonical Forms for Bilinear Forms

Definition 16.12

Suppose that $(-,-) : V \times V \to F$ is a bilinear form. We say that $(-,-)$ is *symmetric* if $(v,w) = (w,v)$ for all $v,w \in V$ and that $(-,-)$ is *skew-symmetric* or *symplectic* if $(v,w) = -(w,v)$ for all $v,w \in V$.

In the main text, we shall only need to deal with bilinear forms that are either symmetric or skew-symmetric. For such a form, $(v,w) = 0$ if and only if $(w,v) = 0$. When $F = \mathbf{R}$, a symmetric bilinear form with $(v,v) \geq 0$ for all $v \in V$ and such that $(v,v) = 0$ if and only if $v = 0$ is said to be an *inner product*.

A vector $v \in V$ is said to be *isotropic* with respect to a form $(-,-)$ if $(v,v) = 0$. For example, if $(-,-)$ is symplectic and the characteristic of the field is not 2, then all elements in V are isotropic. But symmetric bilinear forms can also have isotropic vectors (as long as they do not come from inner products). For example, in Example 16.10 above, the basis of U consists of isotropic vectors.

If $(-,-)$ is non-degenerate and $v \in V$ is isotropic, then there exists some $w \in V$ such that $(v,w) \neq 0$. Clearly v and w must be linearly independent. This observation motivates the following lemma (which we use in Appendix C).

Lemma 16.13

Suppose V has a non-degenerate bilinear form $(-,-)$. Suppose U_1 and U_2 are trivially-intersecting subspaces of V such that $(u,v) = 0$ for all $u,v \in U_1$ and for all $u,v \in U_2$ and that $(-,-)$ restricted to $U_1 \oplus U_2$ is non-degenerate. Then, if $\{u_1, \ldots, u_m\}$ is a basis of U_1 there is a basis $\{u'_1, \ldots, u'_n\}$, of U_2 such that

$$(u_i, u'_j) = \begin{cases} 1 & i = j \\ 0 & i \neq j. \end{cases}$$

Proof

Consider the map $\gamma : U_2 \to U_1^*$ defined by $\gamma(v) = (-, v)$. That is, $\gamma(v)(u) = (u, v)$ for all $u \in U_1$. This map is linear, and it is injective because the restriction of $(-, -)$ to $U_1 \oplus U_2$ is non-degenerate, so we have

$$\dim U_2 \leq \dim U_1^* = \dim U_1.$$

By symmetry, we also have $\dim U_1 \leq \dim U_2$, so γ must be an isomorphism.

Given the basis $\{u_1, \ldots, u_n\}$ of U_1, let $\{f_1, \ldots, f_n\}$ be the corresponding dual basis of U_1^*. For $1 \leq j \leq n$, let $u_j' \in U_2$ be the unique vector such that $\gamma(u_j') = f_j$. Then we have

$$(u_i, u_j') = f_j(u_i) = \begin{cases} 1 & i = j \\ 0 & i \neq j \end{cases}$$

as required. $\square$

Note that if $(-, -)$ is symmetric, then the matrix of $(-, -)$ with respect to this basis of $U_1 \oplus U_2$ is

$$\begin{pmatrix} 0 & I_m \\ I_m & 0 \end{pmatrix}.$$

An analogous result holds if $(-, -)$ is skew-symmetric.

In the following we assume that the characteristic of F is not 2.

Lemma 16.14

Let $(-, -)$ be a non-degenerate symmetric bilinear form on V. Then there is a basis $\{v_1, \ldots, v_n\}$ of V such that $(v_i, v_j) = 0$ if $i \neq j$ and $(v_i, v_i) \neq 0$.

Proof

We use induction on $n = \dim V$. If $n = 1$, then the result is obvious, so we may assume that $\dim V \geq 2$.

Suppose that $(v, v) = 0$ for all $v \in V$. Then, thanks to the identity

$$(v + w, v + w) = (v, v) + (w, w) + 2(v, w),$$

we have $(v, w) = 0$ for all $v, w \in V$, which contradicts our assumption that $(-, -)$ is non-degenerate. (This is where we need our assumption on the characteristic of F.)

We may therefore choose $v \in V$ so that $(v, v) \neq 0$. Let $U = \mathrm{Span}\{v\}$. By hypothesis $U \cap U^\perp = \{0\}$, so by Lemma 16.11 we have $V = U \oplus U^\perp$. Moreover, the restriction of $(-,-)$ to $U^\perp$ is non-degenerate. By the inductive hypothesis, there is a basis of $U^\perp$, say $\{v_2, \ldots, v_n\}$, such that $(v_i, v_j) = 0$ for $i \neq j$ and $(v_i, v_i) \neq 0$ for $2 \leq i \leq n$. Since also $(v, v_j) = 0$ for $j \neq 1$, if we put $v_1 = v$ then the basis $\{v_1, \ldots, v_n\}$ has the required properties. $\square$

Depending on the field, we may be able to be more precise about the diagonal entries $d_i = (v_i, v_i)$. Suppose that $F = \mathbf{R}$. Then we may find $\lambda_i \in \mathbf{R}$ such that $\lambda_i^2 = |d_i|$. By replacing v_i with v_i/λ_i, we may assume that $(v_i, v_i) = \pm 1$. The bilinear form $(-,-)$ is an inner product if and only if $(v_i, v_i) > 0$ for all i.

If $F = \mathbf{C}$, then we can find λ_i so that $\lambda_i^2 = d_i$, and hence we may assume that $(v_i, v_i) = 1$ for all i, so the matrix representing $(-,-)$ is the $n \times n$ identity matrix.

Lemma 16.15

Suppose that $(-,-)$ is a non-degenerate symplectic bilinear form on V. Then we have $\dim V = 2m$ for some m. Moreover, there is a basis of V such that $(v_i, v_{i+n}) \neq 0$ for $1 \leq i \leq n$ and $(v_i, v_j) = 0$ if $|i - j| \neq n$.

Proof

Again we work by induction $\dim V$. Let $0 \neq v \in V$. Since $(-,-)$ is non-degenerate, we may find $w \in V$ such that $(v, w) \neq 0$. Since v, w are isotropic, it is clear that $\{v, w\}$ is linearly independent. Set $v_1 = v$ and $v_2 = w$. If $\dim V = 2$, then we are done. Otherwise, let U be the orthogonal complement of the space W spanned by v_1, v_2. One shows easily that $U \cap W = \{0\}$ and that by dimension counting $V = U \oplus W$. Now, the restriction of $(-,-)$ to U is non-degenerate and also symplectic. The result now follows by induction. $\square$

When $F = \mathbf{R}$ or $F = \mathbf{C}$, it is again useful to scale the basis elements. In particular, when $F = \mathbf{C}$ we may arrange that the matrix representing $(-,-)$ has the form

$$\begin{pmatrix} 0 & I_m \\ -I_m & 0 \end{pmatrix},$$

where I_m is the $m \times m$ identity matrix.

EXERCISES

16.4. Let $x : V \to V$ be a linear transformation of a complex vector space. By the result mentioned at the start of §16.5, we may find a basis

$v_1, \ldots, v_n$ of V in which x is represented by an upper triangular matrix. Let $\lambda_1, \ldots, \lambda_n$ be the diagonal entries of this matrix. Show that if $1 \leq k \leq n$ then

$$(x - \lambda_k 1_V) \ldots (x - \lambda_n 1_V) V \subseteq \mathrm{Span}\{v_1, \ldots, v_{k-1}\}.$$

Hence prove the Cayley–Hamilton theorem for linear maps on complex vector spaces.

16.5. Let A be an $m \times n$ matrix with entries in a field F. Show that there is a bijective correspondence between solution sets of the equation $Ax = y$ for $y \in \mathrm{im}\, A$ and elements of the quotient vector space $F^n / \ker A$.

16.6.† Let V be a finite-dimensional vector space.

(i) Show that $\mathrm{Hom}(V, V)$, the set of linear transformations of V, is a vector space, and determine its dimension.

(ii) Let $A \subseteq \mathrm{Hom}(V, V)$ be a collection of commuting linear maps, each individually diagonalisable. Show that there is a basis of V in which all the elements of A are simultaneously diagonal.

(iii)* Can the assumption that V is finite-dimensional be dropped?

16.7.† Suppose that $x : V \to V$ is a linear map on a vector space V and that $x = d + n = d' + n'$ where d, d' are diagonalisable and n, n' are nilpotent, d and n commute and d' and n' commute. Show that d and d' commute. Hence show that $d - d' = n' - n = 0$. Deduce that the Jordan decomposition of a linear map is unique.

16.8. Let U be a subspace of the F-vector space V.

(i) Consider the restriction map $r : V^\star \to U^\star$, which takes a linear map $f : V \to F$ and regards it just as a map on U. Show that $\ker r = U^\circ$ and $\mathrm{im}\, r = U^\star$. Hence prove that

$$\dim U + \dim U^0 = \dim V.$$

(ii) Now suppose that $(-, -)$ is a non-degenerate bilinear form on V. By considering the linear map $\varphi : V \to U^\star$ defined by

$$\varphi(v)(u) = (u, v),$$

show that $\dim U + \dim U^\perp = \dim V$.

16.9.   Let V be a finite-dimensional vector space with basis $\{v_1, \ldots, v_n\}$. Suppose that $(-, -)$ is a bilinear form on V, and let $a_{ij} = (v_i, v_j)$. Show that $V^\perp = \{0\}$ if and only if the matrix $A = (a_{ij})$ is non-singular.

16.10. Let V be a finite-dimensional vector space and let $\mathrm{Hom}(V, V)$ be the vector space of all linear transformations of V. Show that

$$(x, y) \mapsto \mathrm{tr}(xy)$$

defines a non-degenerate symmetric bilinear form on $\mathrm{Hom}(V, V)$. By Exercise 16.8(ii) this form induces an isomorphism

$$\mathrm{Hom}(V, V) \to \mathrm{Hom}(V, V)^\star.$$

What is the image of the identity map $1_V : V \to V$ under this isomorphism?

Appendix B: Weyl's Theorem

We want to show that any finite-dimensional representation of a complex semisimple Lie algebra is a direct sum of irreducible representations. This is known as Weyl's Theorem; it is a fundamental result in the representation theory of Lie algebras. We used it several times in Chapter 9 to decompose a representation of $\mathsf{sl}(2, \mathbf{C})$ into a direct sum of irreducible representations.

As usual, all Lie algebras and representations in this chapter are finite-dimensional.

17.1 Trace Forms

When we proved that a complex semisimple Lie algebra is a direct sum of simple Lie algebras (or, equivalently, that the adjoint representation is completely reducible), we used the Killing form κ. Now we use a generalisation of the Killing form known as the trace form.

Let L be a Lie algebra and let V be an L-module. Write $\varphi : L \to \mathsf{gl}(V)$ for the corresponding representation. Define the *trace form* $\beta_V : L \times L \to \mathbf{C}$ by

$$\beta_V(x, y) := \operatorname{tr}(\varphi(x) \circ \varphi(y)) \quad \text{for } x, y \in L.$$

This is a symmetric bilinear form on L. In the special case where $V = L$ and φ is the adjoint representation, it is just the Killing form. The trace form β_V is associative; that is,

$$\beta_V([x, y], z) = \beta_V(x, [y, z]) \quad \text{for all } x, y, z \in L,$$

as an easy calculation will show. This implies, as it did for the Killing form, that if we define the radical of β by

$$\operatorname{rad}\beta_V := \{x \in L : \beta_V(x, y) = 0 \text{ for all } y \in L\},$$

then $\operatorname{rad}\beta_V$ is an ideal of L.

So far, our definitions make sense for any representation of an arbitrary Lie algebra. For semisimple Lie algebras, we have the following lemma.

Lemma 17.1

Suppose L is a complex semisimple Lie algebra and $\varphi : L \to \mathsf{gl}(V)$ is a faithful representation (that is, φ is injective). Then $\operatorname{rad}\beta_V$ is zero and so β_V is non-degenerate.

Proof

Let $I = \operatorname{rad}\beta_V$. For any $x, y \in I$, we have $\beta_V(x, y) = 0$. Now apply Proposition 9.3 to the Lie subalgebra $\varphi(I)$ of $\mathsf{gl}(V)$. This gives that $\varphi(I)$ is solvable and hence, since φ is injective, that I is solvable. But L is semisimple, so it has no non-zero solvable ideals, and therefore I is zero.   □

With the assumptions of the previous lemma, we may use the bilinear form β_V to identify L with $L^\star$: namely, given $\theta \in L^\star$, there is a unique element $y \in L$ such that $\beta(x, y) = \theta(x)$ for all $x \in L$. Let $x_1, \ldots, x_n$ be a vector space basis of L and let $\theta_1, \ldots, \theta_n$ be the dual basis of $L^\star$. We use this identification to find elements $y_1, \ldots, y_n$ such that $\beta_V(x, y_j) = \theta_j(x)$ for all $x \in L$, or equivalently such that

$$\beta_V(x_i, y_j) = \begin{cases} 1 & i = j \\ 0 & i \neq j. \end{cases}$$

Lemma 17.2

Suppose that $x \in L$ and $[x_i, x] = \sum_j a_{ij} x_j$. Then, for each t with $1 \leq t \leq n$, we have

$$[x, y_t] = \sum_{i=1}^{n} a_{ti} y_i.$$

Proof

We have

$$\beta_V([x_i, x], y_t) = \sum_{j=1}^{n} a_{ij} \beta_V(x_j, y_t) = a_{it}.$$

We write $[x, y_t] = \sum_{s=1}^{n} b_{ts} y_s$. By associativity,

$$a_{it} = \beta_V([x_i, x], y_t) = \beta_V(x_i, [x, y_t]) = \sum_{s=1}^{n} b_{ts} \beta_V(x_i, y_s) = b_{ti}. \qquad \Box$$

17.2 The Casimir Operator

Let L be a complex semisimple Lie algebra and let V be a faithful L-module with associated representation $\varphi : L \to \mathfrak{gl}(V)$. We continue using the trace form β_V and the two bases of L defined above. The *Casimir operator* associated to φ is the linear map $c : V \to V$ defined by

$$c(v) = \sum_{i=1}^{n} x_i \cdot (y_i \cdot v)$$

in terms of the module action. In the language of representations, c becomes

$$c = \sum_{i=1}^{n} \varphi(x_i) \varphi(y_i).$$

Lemma 17.3

(a) The map $c : V \to V$ is an L-module homomorphism.

(b) We have $\operatorname{tr}(c) = \dim L$.

Proof

For (a), we must show that $c(x \cdot v) - x \cdot (c(v)) = 0$ for all $v \in V$ and $x \in L$, so consider

$$c(x \cdot v) - x \cdot (cv) = \sum_{i=1}^{n} x_i(y_i(xv)) - \sum_{i=1}^{n} x(x_i(y_i v)).$$

Add the equation $-x_i(x(y_i v)) + x_i(x(y_i v)) = 0$ to each term in the sum to get

$$c(x \cdot v) - x \cdot (cv) = \sum_{i=1}^{n} x_i([y_i, x]v) + \sum_{i=1}^{n} [x_i, x](y_i v).$$

If $[x_i, x] = \sum_{j=1}^{n} a_{ij} x_j$, then by the previous lemma we know that $[y_i, x] = \sum_{j=1}^{n} a_{ji} y_j$. Substituting this into the previous equation gives

$$c(x \cdot v) - x \cdot (cv) = \sum_{i,j} -a_{ji} x_i(y_j v) + \sum_{i,j} a_{ij} x_j(y_i v) = 0.$$

For (b), note that

$$\operatorname{tr} c = \sum_{i=1}^{n} \operatorname{tr}(\varphi(x_i) \circ \varphi(y_i)) = \sum_{i=1}^{n} \beta_V(x_i, y_i) = n$$

and $n = \dim L$. $\qquad\qquad\qquad\square$

Theorem 17.4 (Weyl's Theorem)

Let L be a complex semisimple Lie algebra. Every finite-dimensional L-module is completely reducible.

Proof

Let V be such an L-module, and let $\varphi : L \to \mathsf{gl}(V)$ be the corresponding representation. Suppose that W is a submodule of V. By induction on $\dim V$ it is enough to show that W has a complement in V, that is, there is some L-submodule C of V such that $V = W \oplus C$.

We may assume that φ is one-to-one, for if not then we can replace L by the Lie algebra L/I, where I is the kernel of φ. By Lemma 9.12, L/I is semisimple, and if we view V as a module for L/I, then the corresponding representation is one-to-one. It is clear that the submodule structure of V as a module for L/I is the same as the submodule structure of V as a module for L.

We first prove Weyl's Theorem in the special case where $\dim W = \dim V - 1$. In this case, the factor module V/W is the trivial L-module since the derived algebra L' acts trivially on any one-dimensional L-module, and for L semisimple we have $L' = L$ (see Exercise 9.9). Hence

$$x \in L, \ v \in V \implies x \cdot v \in W. \qquad\qquad (\star)$$

We now proceed by induction on $\dim W$.

Step 1: Assume for a start that W is simple. Let $c : V \to V$ be the Casimir operator of V. Since c is an L-module homomorphism, its kernel is a submodule of V. Our aim is to show that $\ker c$ is a suitable complement for W.

As we have noted in $(\star)$, for $x \in L$ and $v \in V$ we have $x \cdot v \in W$. This implies that $c(v) \in W$ for all $v \in V$. In particular, this shows that c is not onto and therefore $\ker c \neq 0$ by the rank-nullity formula.

The restriction of c to W is also an L-module homomorphism, so by Schur's Lemma there is some $\lambda \in \mathbf{C}$ such that $c(w) = \lambda w$ for all $w \in W$. We claim that $\lambda \neq 0$. To see this, we calculate the trace of c in two ways. First, using $c(V) \subseteq W$ and $c(w) = \lambda w$ for $w \in W$, we get that the trace of c is equal to $\lambda \dim W$. On the other hand, from Lemma 17.3 we know that the trace is equal to $\dim L$, which is non-zero. Hence $\lambda \neq 0$. We deduce that $\ker c \cap W = 0$. It now follows from dimension counting that $W \oplus \ker c = V$.

Step 2: For the inductive step, suppose that W_1 is a proper submodule of W. Then W/W_1 is a submodule of V/W_1, and the quotient is 1-dimensional, since by the third isomorphism theorem, we have

$$(V/W_1)/(W/W_1) \cong V/W.$$

Moreover, $\dim V/W_1 < \dim V$, so the inductive hypothesis applied to V/W_1 gives that

$$V/W_1 = W/W_1 \oplus \bar{X},$$

where $\bar{X}$ is a 1-dimensional L-submodule of V/W_1. By the submodule correspondence, there is a submodule X of V containing W_1 such that $\bar{X} = X/W_1$.

Now, $\dim X = 1 + \dim W_1$, and $\dim X < \dim V$ (otherwise we would have $W_1 = W$), so we can also apply the inductive hypothesis to X and get $X = W_1 \oplus C$ for some L-submodule C of X (and again $\dim C = 1$). We claim that $V = W \oplus C$. By dimension counting, it is enough to show that $W \cap C = 0$. The direct sum decomposition of V/W_1 above implies that the image of $W \cap X$ under the quotient map $V \to V/W_1$ is zero. Hence $W \cap X \subseteq W_1$ and so $W \cap C \subseteq W_1 \cap C = 0$.

This completes the proof of Weyl's Theorem in the special case where $\dim W + 1 = \dim V$.

We now consider the general case. Suppose W is an L-submodule of V. Consider the L-module $M := \mathrm{Hom}(V, W)$ of linear maps from V to W. Recall from Exercise 7.12 that the action of L on M is defined by

$$(x \cdot f)(v) = x \cdot f(v) - f(x \cdot v) \quad \text{for } x \in L, f \in M, v \in V.$$

We can apply the result for the special case to M as follows. Let

$$M_S := \{f \in M : f{\downarrow}_W = \lambda 1_W \text{ for some } \lambda \in \mathbf{C}\},$$
$$M_0 := \{f \in M : f{\downarrow}_W = 0\},$$

where $f{\downarrow}_W$ denotes the restricted map $f : W \to W$. One can check that both M_S and M_0 are L-submodules of M, and $M_0 \subseteq M_S$.

We claim that the quotient M_S/M_0 is 1-dimensional. Clearly, the identity map 1_V of V lies in M_S but not in M_0, so the coset of 1_V is a non-zero element in the quotient. Now if $f \in M_S$ satisfies $f{\downarrow}_W = \lambda 1_W$, then $f - \lambda 1_V$ belongs to M_0; that is, $f + M_0 = \lambda 1_V + M_0$.

We now apply Weyl's Theorem for the special case. This tells us that $M_S = M_0 \oplus C$ for some L-submodule C of M_S. Also, C is 1-dimensional and hence trivial as an L-module, so C contains a non-zero element φ such that $x \cdot \varphi = 0$ for all $x \in L$. The condition $x \cdot \varphi = 0$ just means that φ is an L-module homomorphism. By scaling φ we may assume that $\varphi{\downarrow}_W = 1_W$.

Now we can get back to the original module V. Since φ is an L-module homomorphism, its kernel, say K, is a submodule of V. We claim that $V = K \oplus W$. If $v \in K \cap W$, then $\varphi(v) = 0$. On the other hand, the restriction of φ to W is the identity, so $\varphi(v) = v$. Therefore $K \cap W = 0$. By the definition of M, $\operatorname{im} \varphi$ is contained in W. It now follows from the rank-nullity formula that

$$\dim K = \dim V - \dim \operatorname{im} \varphi \geq \dim V - \dim W,$$

so by dimension counting $V = K + W$. Hence $V = K \oplus W$, as required. $\square$

EXERCISES

17.1. Let V be a representation of the complex semisimple Lie algebra L. Prove that the Casimir operator $c : V \to V$ is independent of the choice of basis of L.

17.2. Show that the Casimir operator for the natural representation of $\mathsf{sl}(2, \mathbf{C})$ with respect to the standard basis (h, e, f) of $\mathsf{sl}(2, \mathbf{C})$ is given by

$$c(v) = (ef + fe + \tfrac{1}{2}h^2)v.$$

Show that if V is any irreducible representation of $\mathsf{sl}(2, \mathbf{C})$, then the Casimir operator for V is given by a scalar multiple of this expression.

17.3. Prove that a complex Lie algebra is semisimple if and only if all its finite-dimensional representations are completely reducible.

18

Appendix C: Cartan Subalgebras

Suppose that L is a semisimple Lie algebra with Cartan subalgebra H and associated root system Φ. We want to show that if H_1 is another Cartan subalgebra of L, with associated root system Φ_1, then Φ_1 is isomorphic to Φ. This shows first of all that the root system of a semisimple Lie algebra is well-defined (up to isomorphism) and secondly that semisimple Lie algebras with different root systems cannot be isomorphic.

The general proof of this statement is quite long and difficult and requires several ideas which we have so far avoided introducing. So instead we give a proof that assumes that L is a classical Lie algebra. This will be sufficient to show that the only isomorphisms between the classical Lie algebras come from isomorphisms between their root systems; we used this fact at the end of Chapter 12. We then show how Serre's Theorem and the classification of Chapter 14 can be used to give the result for a general semisimple Lie algebra.

We conclude by discussing the connection between our Cartan subalgebras and the "maximal toral algebras" used by other authors, such as Humphreys [14].

18.1 Root Systems of Classical Lie Algebras

Let L be a classical Lie algebra, so L is defined as a particular subalgebra of a matrix algebra $\mathsf{gl}(n, \mathbf{C})$. This gives us a practical way to show that the root systems corresponding to different Cartan subalgebras are isomorphic.

Proposition 18.1

Let H and K be Cartan subalgebras of L. Suppose that there is an invertible $n \times n$ matrix P such that $PHP^{-1} \subseteq K$ and $PLP^{-1} \subseteq L$. Then the root systems corresponding to H and K are isomorphic.

Proof

We first show that the hypotheses imply $PHP^{-1} = K$ and $PLP^{-1} = L$. The latter follows just by dimension counting. For the former, we argue that because an element of L is semisimple if and only if it is diagonalisable (see Theorem 9.16) and, moreover, a conjugate of a diagonalisable matrix is diagonalisable, the elements of $P^{-1}KP$ are semisimple. Hence $P^{-1}KP$ is an abelian subalgebra of L consisting of semisimple elements and containing H. Since H is a Cartan subalgebra, we must have $H = P^{-1}KP$.

Now let α be a root taken with respect to H, and let $x \in L$ be in the corresponding root space, so $[h, x] = \alpha(h)x$ for all $h \in H$. Then

$$[PhP^{-1}, PxP^{-1}] = P[h, x]P^{-1} = \alpha(h)PxP^{-1}.$$

Hence, if we define $\alpha^P \in K^\star$ by

$$\alpha^P(k) = \alpha(P^{-1}kP) \quad \text{for } k \in K,$$

then PxP^{-1} is in the α^P root space of K. Therefore $\alpha \mapsto \alpha^P$ is a one-to-one map from the roots of H to the roots of K; by symmetry it is a bijection.

In fact, this map induces an isomorphism between the root systems corresponding to H and K. To check condition (b) in the definition of isomorphism (Definition 11.19), note that if α and β are roots with respect to H, then

$$\langle \alpha^P, \beta^P \rangle = \alpha^P(h_{\beta^P}) = \alpha(P^{-1}h_{\beta^P}P) = \alpha(h_\beta) = \langle \alpha, \beta \rangle.$$

Here we used that $h_{\alpha^P} = Ph_\alpha P^{-1}$; one way to prove this is to consider the action of $Ph_\alpha P^{-1}$ on the K-root spaces where K acts as $\pm \alpha^P$. □

For the classical Lie algebras, we have a standard choice of Cartan subalgebra, namely the subalgebra consisting of all diagonal matrices in the Lie algebra. So given a Cartan subalgebra H, we look for a basis of $V = \mathbf{C}^n$ in which the elements of H are diagonal. We can then apply Proposition 18.1 with P as the change of basis matrix.

We start with the Lie algebras $\mathsf{sl}(n, \mathbf{C})$ for $n \geq 2$. In this case, because commuting diagonalisable matrices may be simultaneously diagonalised, and conjugation preserves traces, the hypotheses for Proposition 18.1 are easily seen to be satisfied.

18.2 Orthogonal and Symplectic Lie Algebras

We now consider $\mathsf{sp}(2m, \mathbf{C})$ for $m \geq 1$ and $\mathsf{so}(n, \mathbf{C})$ for $n \geq 3$. Note that we do not consider

$$\mathsf{so}(2, \mathbf{C}) = \left\{ \begin{pmatrix} c & 0 \\ 0 & -c \end{pmatrix} : c \in \mathbf{C} \right\}$$

as it is abelian and so (by definition) not semisimple.

We defined the algebras $\mathsf{so}(2\ell, \mathbf{C}), \mathsf{so}(2\ell + 1, \mathbf{C}), \mathsf{sp}(2\ell, C)$ in §4.3 as the set of matrices $\{x \in \mathsf{gl}(n, \mathbf{C}) : x^t S + Sx = 0\}$, where S is

$$\begin{pmatrix} 0 & I_\ell \\ I_\ell & 0 \end{pmatrix}, \quad \begin{pmatrix} 1 & 0 & 0 \\ 0 & 0 & I_\ell \\ 0 & I_\ell & 0 \end{pmatrix}, \quad \begin{pmatrix} 0 & I_\ell \\ -I_\ell & 0 \end{pmatrix},$$

respectively. Since we shall need to change bases, it is advantageous to translate the defining condition into the language of bilinear forms. We may define a nonsingular bilinear form on $V = \mathbf{C}^n$ by setting

$$(v, w) := v^t S w \quad \text{for all } v, w \in V.$$

With this convention, each family of classical Lie algebras may be defined by

$$\{x \in \mathsf{gl}(V) : (xv, w) = -(v, xw) \text{ for all } v, w \in V\}.$$

The matrix S may be recovered as the matrix of the form $(-, -)$ with respect to the standard basis $\varepsilon_1, \ldots, \varepsilon_n$ of V.

By Theorem 10.4, there exists $h \in H$ such that $H = C_L(h)$. As an element of a Cartan subalgebra, h is semisimple. It follows that h acts diagonalisably on V, and so we may write V as a direct sum of eigenspaces for h. Let

$$V = \bigoplus_{\lambda \in \Psi} V_\lambda,$$

where Ψ is the set of eigenvalues of h and V_λ is the eigenspace for eigenvalue λ.

Lemma 18.2

(a) Suppose that $\lambda, \mu \in \Psi$ and $\lambda + \mu \neq 0$. Then $V_\lambda \perp V_\mu$. In particular, $(v, v) = 0$ for all $v \in V_\lambda$ when $\lambda \neq 0$.

(b) If $\lambda \in \Psi$, then $-\lambda \in \Psi$.

(c) If $\lambda \in \Psi$ is non-zero, then the restriction of the bilinear form to $V_\lambda \oplus V_{-\lambda}$ is nonsingular.

(d) If $0 \in \Psi$, then the bilinear form restricted to V_0 is nonsingular.

Proof

We adapt the proof of Lemma 10.1. For (a), note that if $v \in V_\lambda$ and $w \in V_\mu$, then

$$\lambda(v, w) = (\lambda v, w) = (hv, w) = -(v, hw) = -\mu(v, w).$$

Hence $(\lambda + \mu)(v, w) = 0$ and so $(v, w) = 0$. For (b), take $0 \neq v \in V_\lambda$; since $(-, -)$ is non-degenerate, there is some $x \in V$ such that $(v, x) \neq 0$. Write $x = \sum x_\mu$, where $x_\mu \in V_\mu$. Then

$$0 \neq (v, x) = \sum_\mu (v, x_\mu) = (v, x_{-\lambda})$$

and hence $V_{-\lambda}$ must be non-zero. Parts (c) and (d) follow easily from part (a) and the non-degeneracy of $(-, -)$. $\square$

Let λ be a non-zero weight of H in its action on V. By part (a) of the previous lemma, $(u, v) = 0$ for all $u, v \in V_\lambda$ and also for all $u, v \in V_{-\lambda}$. By part (c), the restriction of $(-, -)$ to $V_\lambda \oplus V_{-\lambda}$ is non-degenerate, so it follows from Lemma 16.13 that for any basis $\{f_i\}$ of V_λ there is a vector space basis $\{f_i'\}$ of $V_{-\lambda}$ such that

$$(f_i, f_j') = \begin{cases} 1 & i = j \\ 0 & i \neq j. \end{cases}$$

The matrix describing the bilinear form on $V_\lambda \oplus V_{-\lambda}$ with respect to this basis is therefore one of

$$\begin{pmatrix} 0 & I \\ I & 0 \end{pmatrix}, \quad \begin{pmatrix} 0 & I \\ -I & 0 \end{pmatrix},$$

depending on whether $(-, -)$ is symmetric (which is the case when L is orthogonal) or skew-symmetric (which is the case when L is symplectic).

Now suppose h has a zero eigenvalue. By part (d) of the previous lemma, the restriction of the bilinear form to V_0 is non-degenerate. Hence we can choose a basis of V_0 so that the matrix of the bilinear form $(-, -)$ restricted to V_0 is of the same form as the original matrix S.

We can put together the bases just constructed to get a basis of V, say $\mathcal{B} = \{b_1, \ldots, b_n\}$. This basis consists of eigenvectors of h. Moreover, the matrix of the bilinear form $(-, -)$ with respect to $\mathcal{B}$ is the same as its matrix with respect to the standard basis, namely S.

Proposition 18.3

With respect to the basis $\mathcal{B}$, the matrices of elements of H are diagonal.

Proof

We write a general $x \in H = C_L(h)$ with respect to the basis $\mathcal{B}$. Since x commutes with h, it preserves the eigenspaces of h. It follows that the matrix of x restricted to $V_\lambda \oplus V_{-\lambda}$ must be of the form

$$\begin{pmatrix} a & 0 \\ 0 & b \end{pmatrix}$$

for some matrices a and b of size $\dim V_\lambda$. The matrix with respect to $\mathcal{B}$ of the bilinear form $(-,-)$ on $V_\lambda \oplus V_{-\lambda}$ is $\begin{pmatrix} 0 & I \\ \pm I & 0 \end{pmatrix}$; hence

$$x^t \begin{pmatrix} 0 & I \\ \pm I & 0 \end{pmatrix} + \begin{pmatrix} 0 & I \\ \pm I & 0 \end{pmatrix} x = 0.$$

Therefore the matrix of x with respect to $\mathcal{B}$ is

$$\begin{pmatrix} a & 0 \\ 0 & -a^t \end{pmatrix}$$

where a can be any matrix of size $\dim V_\lambda$. As H is abelian we must have $m = 1$ (otherwise there are non-commuting matrices of this form).

Now consider the block of x corresponding to V_0. Let $m = \dim V_0$. If our Lie algebra is $\mathsf{sp}_{2m}(\mathbf{C})$ then we must have $x \in \mathsf{sp}_{2m}(\mathbf{C})$. Since $\mathsf{sp}_2(\mathbf{C}) = \mathsf{sl}_2(\mathbf{C})$ is non-abelian, this forces $m = 0$. On the other hand, if we are considering $\mathsf{so}_n(\mathbf{C})$, then we may have $m = 2$ since, as noted earlier, $\mathsf{so}_2(\mathbf{C})$ already consists of diagonal matrices, but $m > 2$ is impossible since $\mathsf{so}_3(\mathbf{C})$ is non-abelian.

Therefore, in either case, the matrix of x with respect to our chosen basis of V is diagonal. $\qquad\square$

Corollary 18.4

Define $P : V \to V$ by $Pb_i = \varepsilon_i$, where $\varepsilon_1, \ldots, \varepsilon_n$ is the standard basis of $V = \mathbf{C}^n$. Then every element of PHP^{-1} is diagonal and $PLP^{-1} \subseteq L$.

Proof

We first note that $P^{-t}SP^{-1}$ is the matrix of the bilinear form $(-,-)$ with respect to the basis $\mathcal{B}$. By construction, this is the same as S, so $P^{-t}SP^{-1} = S$, which gives $S = P^t S P$.

Now, to show that $PLP^{-1} = L$, it is sufficient to prove that for each $x \in L$

$$(PxP^{-1})^t S + S(PxP^{-1}) = 0.$$

This is easily seen to be equivalent to

$$x^t P^t SP + P^t SPx = 0,$$

which holds by the first paragraph and the definition of L.

Finally, for each $h \in H$, we have $PhP^{-1}\varepsilon_i = PhP^{-1}Pb_i = Phb_i \in$ Span$\{Pb_i\}$ = Span$\{\varepsilon_i\}$, so PHP^{-1} consists of diagonal matrices. The maximality of H now guarantees that every diagonal matrix in L must appear. □

Hence, by Proposition 18.1, the root system corresponding to H is isomorphic to the root system corresponding to the Cartan subalgebra of diagonal matrices.

18.3 Exceptional Lie Algebras

Lemma 18.5

Let L be an exceptional simple Lie algebra with Cartan subalgebra H and root system Φ. If H_1 is another Cartan subalgebra of L, then its root system Φ_1 is isomorphic to Φ.

Proof

We use the classification of root systems in Chapter 13. If Φ_1 has type A, B, C, or D, then Serre's Theorem implies that L is isomorphic to a classical Lie algebra. So we would have a classical Lie algebra with a root system of type E, F, or G, in contradiction to the previous section.

Therefore Φ_1 must be an exceptional root system, and so by Serre's theorem we get an isomorphism between two exceptional Lie algebras. But, by Exercise 10.5, the dimensions of the exceptional Lie algebras can be found from their root systems to be

type	G_2	F_4	E_6	E_7	E_8
dimension	14	52	78	133	248

so this is only possible if Φ_1 is isomorphic to Φ, as required. □

18.4 Maximal Toral Subalgebras

A Lie algebra is said to be *toral* if it consists entirely of semisimple elements. A *maximal toral subalgebra* of a Lie algebra L is a Lie subalgebra of L which is toral and which is not contained in any larger toral subalgebra. Our aim is to show that the maximal toral subalgebras of L are precisely the Cartan subalgebras.

Lemma 18.6

If L is a semisimple Lie algebra and T is a toral subalgebra of L, then T is abelian.

Proof

Take $s \in T$. As $\operatorname{ad} s$ acts diagonalisably on L and preserves T, the restriction of $\operatorname{ad} s$ to T is also diagonalisable. We must show that it only has zero as an eigenvalue.

Suppose that there is some non-zero $t \in T$ such that $(\operatorname{ad} s)t = ct$ with $c \neq 0$. We shall obtain a contradiction by rewriting this as $(\operatorname{ad} t)s = -ct$. As $\operatorname{ad} t$ also acts diagonalisably on T, we may extend t to a basis of T consisting of eigenvectors for $\operatorname{ad} t$, say $\{t, y_1, \ldots, y_{m-1}\}$. Let

$$s = \lambda t + \mu_1 y_1 + \ldots + \mu_m y_m.$$

Then

$$-ct = (\operatorname{ad} t)s = \mu_1 (\operatorname{ad} t)y_1 + \ldots + \mu_m (\operatorname{ad} t)y_m$$

is a non-trivial linear dependency between elements of our chosen basis of T, so we have reached a contradiction. $\square$

Corollary 18.7

A subalgebra of L is maximal toral if and only if it is a Cartan subalgebra.

Proof

Since the previous lemma shows that toral subalgebras are always abelian, a maximal toral subalgebra is the same thing as a maximal abelian toral subalgebra; that is, a Cartan subalgebra. $\square$

Appendix D: Weyl Groups

Our first aim in this appendix is to prove the following theorem.

Theorem 19.1

Let R be a root system with base B and Weyl group W. If $w \in W$, then $w(B)$ is also a base of R. Moreover, if B' is another base of R, then there exists a unique $w \in W$ such that $w(B) = B'$.

We use part of this theorem in the main text to show that the Cartan matrix and Dynkin diagram of a root system do not depend on the choice of base.

We then describe the structure of the Weyl groups of root systems of types A, B, C, and D.

19.1 Proof of Existence

The first part of the theorem is very easy to prove. Let $B = \{\alpha_1, \ldots, \alpha_\ell\}$ and let α be a root. Then $w^{-1}\alpha$ is also a root (since the Weyl group permutes the roots). Suppose that

$$w^{-1}\alpha = \sum_{i=1}^{\ell} k_i \alpha_i,$$

where the coefficients k_i all have the same sign. Then

$$\alpha = \sum_{i=1}^{\ell} k_i w(\alpha_i)$$

with the same coefficients and hence $w(B)$ is a base. This shows that the Weyl group permutes the collection of bases of R.

We now prove that the Weyl group acts transitively. Let R^+ denote the set of positive roots of R with respect to B. Let $B' = \{\alpha'_1, \ldots, \alpha'_\ell\}$ be another base of R. Let R'^+ denote the set of positive roots of R, taken with respect to B', and let R'^- denote the set of negative roots of R, again taken with respect to B'. Note that all these sets have the same size, namely half the size of R.

The proof will be by induction on $|R^+ \cap R'^-|$. Interestingly, the base case, where $R^+ \cap R'^- = \emptyset$, is the hardest part. In this case, we must have $R^+ = R'^+$, so the bases B and B' give the same positive roots. Each element of B' is a positive root with respect to B, so we may define a matrix P by

$$\alpha'_j = \sum_{i=1}^{\ell} p_{ij} \alpha_i,$$

whose coefficients are all non-negative integers. Similarly we may define a matrix Q by

$$\alpha_k = \sum_{j=1}^{\ell} q_{jk} \alpha'_j.$$

These matrices have the property that $PQ = QP = I$. Now, by Exercise 19.1 below, this can only happen if P and Q are permutation matrices. Hence the sets B and B' coincide, and so for the element w we may take the identity.

Now suppose that $|R^+ \cap R'^-| = n > 0$. Then $B \cap R'^- \neq \emptyset$ since otherwise $B \subseteq R'^+$, which implies $R^+ \subseteq R'^+$ and hence $R^+ = R'^+$, since both sets have the same size. Take some $\alpha \in B \cap R'^-$, and let $s = s_\alpha$. Then $s(R^+)$ is the set of roots obtained from R^+ by replacing α by $-\alpha$, so the intersection $s(R^+) \cap R'^-$ has $n-1$ elements. The set $s(R^+)$ is the set of positive roots with respect to the base $s(B)$. By the inductive hypothesis, there is some $w_1 \in W$ such that $w_1(s(B)) = B'$. Now take $w = w_1 s$; this sends B to B'. $\qquad\square$

We have already proved enough of the theorem for our applications in the main text. It remains to show that the element $w \in W$ which we have found is unique.

19.2 Proof of Uniqueness

We keep the notation of the previous section.

Lemma 19.2

Let $\alpha_1, \alpha_2, \ldots, \alpha_t \in B$ be not necessarily distinct simple roots and set $s_i = s_{\alpha_i} \in W$. Suppose that $s_1 s_2 \ldots s_{t-1}(\alpha_t) \in R^-$. Then there is some r such that $1 \le r < t$ and

$$s_1 s_2 \ldots s_t = (s_1 s_2 \ldots s_{r-1})(s_{r+1} \ldots s_{t-1}).$$

Proof

Let $\beta_i = s_{i+1} \ldots s_{t-1}(\alpha_t)$ for $0 \le i \le t-2$ and let $\beta_{t-1} = \alpha_t$. Then $\beta_0 \in R^-$ by assumption, and $\beta_{t-1} \in R^+$ since α_t is a simple root. Hence there is a smallest r with $1 \le r \le t-1$ such that $\beta_r \in R^+$ and $s_r(\beta_r) \in R^-$. By Lemma 11.13, the only positive root sent by the simple reflection s_r to a negative root is α_r, so we must have $\beta_r = \alpha_r$.

Therefore we have $\alpha_r = w(\alpha_t)$, where $w = s_{r+1} \ldots s_{t-1}$. Now we analyse what this means for the reflection s_r. Substituting, we get

$$s_r = s_{\alpha_r} = s_{\beta_r} = s_{w\alpha_t}.$$

But we know from Exercise 11.6 that $s_{w(\alpha_t)} = w s_{\alpha_t} w^{-1}$. Since w^{-1} is just the product of the reflections making up w taken in the opposite order, we get

$$s_r = s_{r+1} \ldots s_{t-1} s_t s_{t-1} \ldots s_{r+1}.$$

We substitute this and get

$$s_1 \ldots s_r \ldots s_t = s_1 \ldots s_{r-1} \left(s_{r+1} \ldots s_{t-1} s_t s_{t-1} \ldots s_{r+1} \right) s_{r+1} \ldots s_{t-1} s_t,$$

and cancelling gives the stated answer. $\qquad\square$

Corollary 19.3

Suppose $w = s_1 \ldots s_t \in W$ with $s_i = s_{\alpha_i}$ and $\alpha_i \in B$. If t is minimal such that w is a product of this form, then w takes α_t to a negative root.

Proof

We have $w(\alpha_t) = -s_1 \ldots s_{t-1}(\alpha_t)$. Suppose, for a contradiction, that this is a positive root. Then $s_1 \ldots s_{t-1}(\alpha_t)$ is a negative root, so by the previous lemma we can write $s_1 \ldots s_{t-1}$ as a product of fewer reflections. It follows also that w can be written as a product of fewer than t simple reflections, which contradicts the hypothesis. $\qquad\square$

We can now prove uniqueness. It is sufficient to prove that if $w \in W$ and $w(B) = B$, then w is the identity. Suppose not, and write w as a product of reflections $w = s_1 \ldots s_t$, where $s_i = s_{\alpha_i}$ for $\alpha_i \in B$ and t is as small as possible. By assumption, $t \geq 1$, so by the previous corollary $w(\alpha_t)$ is a negative root and hence it is not in B, a contradiction to the assumption that $w(B) = B$. This shows that $w = 1$.

19.3 Weyl Groups

We now discuss the Weyl groups of the root systems of types A, B, C, and D using the constructions given in Chapter 12. We need to calculate the action of the simple reflections s_α for α in a base of the root system. Since we chose bases containing as many of the elements

$$\alpha_i := \varepsilon_i - \varepsilon_{i+1}$$

as possible, it will be useful to calculate the action of the reflection s_{α_i} on $\mathbf{R}^m$. We have

$$
\begin{aligned}
s_{\alpha_i}(v) &= v - \langle v, \alpha_i \rangle \alpha_i \\
&= (v_1, \ldots, v_m) - (v_i - v_{i+1})(\varepsilon_i - \varepsilon_{i+1}) \\
&= (v_1, \ldots, v_{i-1}, v_{i+1}, v_i, v_{i+2}, \ldots, v_m).
\end{aligned}
$$

Thus, s_{α_i} permutes the i-th and $(i+1)$-th coordinate of vectors.

19.3.1 The Weyl Group of Type A_ℓ

The base given in §12.2 is $\alpha_1, \ldots, \alpha_\ell$, so the Weyl group is generated by the reflections s_{α_i}, whose action we just calculated. More generally, this calculation shows that the reflection $s_{\varepsilon_i - \varepsilon_j}$ swaps the i-th and j-th coordinates of a vector.

We can use this to identify the group structure of the Weyl group; namely, W acts by linear transformations on $\mathbf{R}^{\ell+1}$, and in this action it permutes the standard basis vectors ε_i. For example, the reflection s_{α_i} swaps ε_i and ε_{i+1} and fixes the other basis vectors, so we get a group homomorphism $\rho : W \to \mathcal{S}_{\ell+1}$, where $\rho(w)$ is the permutation induced on the standard basis of $\mathbf{R}^{\ell+1}$. The symmetric group is generated by transpositions so ρ is onto. Furthermore, ρ is injective, for if w fixes every ε_i, then w must be the identity map, so W is isomorphic to the symmetric group $\mathcal{S}_{\ell+1}$.

19.3.2 The Weyl Group of Type B_ℓ

The base constructed in §12.3 is $\alpha_1, \ldots, \alpha_{\ell-1}, \beta_\ell$, where $\beta_\ell = \varepsilon_\ell$. We start by looking at the simple reflections. For $1 \le i < \ell$ we have already calculated their action at the start of §13.2. To find s_{β_ℓ}, we note that for every $v \in \mathbf{R}^\ell$ we have $\langle v, \beta_\ell \rangle = 2v_\ell$ and hence

$$s_{\beta_\ell}(v_1, \ldots, v_\ell) = (v_1, \ldots, -v_\ell).$$

Similarly, for any i, we have $\langle v, \varepsilon_i \rangle = 2v_i$ and hence

$$s_{\varepsilon_i}(v_1, \ldots, v_\ell) = (v_1, \ldots, -v_i, \ldots, v_\ell).$$

Let N be the subgroup of W generated by $s_{\varepsilon_1}, \ldots, s_{\varepsilon_\ell}$. The generators satisfy $s_{\varepsilon_i}^2 = 1$ and they commute, so N is isomorphic to a direct product of ℓ copies of the cyclic group of order 2.

We claim that N is a normal subgroup of W. To see this, one checks that s_{α_i} and s_{ε_j} commute for $j \ne i, i+1$, and that

$$s_{\alpha_i} s_{\varepsilon_{i+1}} s_{\alpha_i} = s_{\varepsilon_i}$$

and hence $s_{\alpha_i} s_{\varepsilon_i} s_{\alpha_i} = s_{\varepsilon_{i+1}}$. (Recall that s_{α_i} is its own inverse.) Since the s_{α_i} and s_{β_ℓ} generate W and $s_{\beta_n} \in N$, this is sufficient to show that N is normal.

Finally, we claim that the factor group W/N is isomorphic to the symmetric group $\mathcal{S}_\ell$. We have seen that in the conjugacy action of W on N, the generators $\{s_{\varepsilon_1}, \ldots, s_{\varepsilon_\ell}\}$ are permuted. This gives us a homomorphism $W \to \mathcal{S}_\ell$. As N is abelian, the kernel of this homomorphism contains N.

Let S denote the subgroup of W generated by $s_{\alpha_1}, \ldots, s_{\alpha_{\ell-1}}$. We know that S is isomorphic to $\mathcal{S}_l$ from §19.3.1. It is easy to see by looking at the action of S and N on $\pm\varepsilon_1, \ldots, \pm\varepsilon_\ell$ that $S \cap N = \{1\}$. Hence we have $W/N = SN/N \cong S/S \cap N \cong \mathcal{S}_\ell$. More concretely, we may observe that the simple reflection s_{α_i} acts as the transposition $(s_{\varepsilon_i} s_{\varepsilon_{i+1}})$ on N, so the action of $W/N = \langle s_{\alpha_1} N, \ldots, s_{\alpha_{n-1}} N \rangle$ on the generators of N is equivalent to the action of $\mathcal{S}_\ell$ on $\{1 \ldots n\}$.

19.3.3 The Weyl Group of Type C_ℓ

It will be seen that the only difference between the root systems of type B_ℓ and C_ℓ we constructed in Chapter 12 is in the relative lengths of the roots. This does not affect the reflection maps: If $v \in \mathbf{R}^\ell$, then s_v is the same linear map as $s_{\lambda v}$ for any $\lambda \in \mathbf{R}$, so the Weyl group of type C_ℓ is the same as that of type B_ℓ. (Underlying this is a duality between B_ℓ and C_ℓ — see Exercise 13.2 below.)

19.3.4 The Weyl Group of Type D_ℓ

Here our base was $\alpha_1, \ldots, \alpha_{\ell-1}, \beta_\ell$, where $\beta_\ell = \varepsilon_{\ell-1} + \varepsilon_\ell$. We find that $s_{\beta_\ell}(v_1, \ldots, v_{\ell-1}, v_\ell) = (v_1, \ldots, -v_\ell, -v_{\ell-1})$. Therefore, the composite map $s_{\alpha_{\ell-1}} s_{\beta_\ell}$ acts as

$$s_{\alpha_{n-1}} s_{\beta_n}(v_1, \ldots, v_{\ell-1}, v_\ell) = (v_1, \ldots, -v_{\ell-1}, v_\ell).$$

More generally, if we set $\beta_i = \varepsilon_{i-1} + \varepsilon_i$ and $t_{i-1} := s_{\alpha_{i-1}} s_{\beta_i}$, then

$$t_{i-1}(v) = (v_1, \ldots, -v_{i-1}, -v_i, v_{i+1}, \ldots, v_\ell).$$

Let N be the subgroup generated by $t_1, \ldots, t_{\ell-1}$. The generators commute and square to the identity, so N is isomorphic to a direct product of $\ell - 1$ copies of the cyclic group of order 2. In fact, N is normal in W and the quotient W/N is isomorphic to the symmetric group $\mathcal{S}_\ell$.

EXERCISES

19.1.† Suppose that P and Q are matrices, all of whose entries are non-negative integers. Show that if $PQ = I$ then P and Q are permutation matrices. That is, each row and column of P has a unique non-zero entry, and this entry is a 1. For instance,

$$\begin{pmatrix} 1 & 0 & 0 \\ 0 & 0 & 1 \\ 0 & 1 & 0 \end{pmatrix}$$

is a permutation matrix corresponding to the transposition (23).

19.2.* This exercise gives an alternative, more geometric proof that the Weyl group of a root system acts transitively on its collection of bases. For solutions, see Chapter 10 of Humphreys, *Introduction to Lie Algebras and Representation Theory* [14].

(i) Let V be a real inner-product space, and let $v_1, \ldots, v_k$ be vectors in V. Show that there exists $z \in V$ such that $(z, v_i) > 0$ for $1 \leq i \leq k$ if and only if $v_1, \ldots, v_k$ are linearly independent. *Hint*: For the "if" direction, let $V_j = \mathrm{Span}\{v_i : i \neq j\}$, take $w_j \in V_j^\perp$, and consider a vector of the form $\sum_j c_j v_j$ for scalars $c_j \in \mathbf{R}$. For the "only if" direction see the proof of Theorem 11.10.

(ii) Let R be a root system in the real inner-product space E.

(a) Show that if $B = \{\alpha_1, \ldots, \alpha_\ell\}$ is a base for a root system R, then there exists $z \in E$ such that $(\alpha_i, z) > 0$ for $1 \le i \le \ell$.

(b) Show that $R^+ = \{\alpha \in R : (\alpha, z) > 0\}$.

(c) Say that $\beta \in R^+$ is *indecomposable* if it cannot be written as $\beta_1 + \beta_2$ for $\beta_1, \beta_2 \in R^+$. Show that B is the set of indecomposable elements of R^+. Thus the construction used to prove Theorem 11.10 in fact gives every base of R.

(d) Suppose $v \in E$ and $(\alpha, v) \ne 0$ for all $\alpha \in R$. Prove that there exists $w \in W$ such that $(\alpha, w(v)) > 0$ for all $\alpha \in R^+$. *Hint:* Let $\delta = \frac{1}{2} \sum_{\alpha \in R^+} \alpha$, and choose $w \in W$ so that $(w(\alpha), \delta)$ is maximised.

(e) Prove that if B' is another base of W, then there exists $w \in W$ such that $w(B') = B$.

19.3.   Prove the statements made about the Weyl group of type D_n. (*Hint:* Mimic the proof for type B_n.)

20
Appendix E: Answers to Selected Exercises

If an exercise is used later in the main text, then we have usually provided a solution, in order to prevent the reader from getting needlessly stuck. We do, however, encourage the reader to make a thorough unaided attempt before evaluating what we have to offer. There are usually several good solutions, and yours might well be just as good as ours, or even better!

Chapter 1

1.11. An isomorphism of L_1 and L_2 is necessarily an isomorphism of their underlying vector spaces, so if L_1 and L_2 are isomorphic, then they have the same dimension.

Conversely, if L_1 and L_2 have the same dimension, then there is an invertible linear map $f : L_1 \to L_2$. For $x, y \in L_1$, we have

$$f([x, y]) = 0 = [f(x), f(y)],$$

so f is also an isomorphism of Lie algebras.

1.13. The structure constants are determined by the Lie brackets $[h, e] = 2e$, $[h, f] = -2f$, and $[e, f] = h$.

1.14. (i) The vector space of 3×3 antisymmetric matrices is 3-dimensional.

One possible basis is given by the matrices

$$X = \begin{pmatrix} 0 & 1 & 0 \\ -1 & 0 & 0 \\ 0 & 0 & 0 \end{pmatrix}, \ Y = \begin{pmatrix} 0 & 0 & 0 \\ 0 & 0 & 1 \\ 0 & -1 & 0 \end{pmatrix}, \ Z = \begin{pmatrix} 0 & 0 & 1 \\ 0 & 0 & 0 \\ -1 & 0 & 0 \end{pmatrix}.$$

This basis has been chosen so that $[X, Y] = Z$, $[Y, Z] = X$, $[Z, X] = Y$. Hence the linear map $\theta : L \to \mathsf{gl}(3, \mathbf{C})$ defined on basis elements by $\theta(x) = X$, $\theta(y) = Y$, and $\theta(z) = Z$ gives one possible isomorphism.

(ii) Let (e, f, h) be the basis of $\mathsf{sl}(2, \mathbf{C})$ given in Exercise 1.13. We look for a possible image of h under a Lie algebra isomorphism $\varphi : \mathsf{sl}(2, \mathbf{C}) \to L$.

Let $\alpha, \beta, \gamma \in \mathbf{C}$ and let $u = \alpha x + \beta y + \gamma z \in L$. The matrix of ad u is

$$\begin{pmatrix} 0 & -\gamma & \beta \\ \gamma & 0 & -\alpha \\ -\beta & \alpha & 0 \end{pmatrix},$$

which has characteristic polynomial

$$\chi(X) = -X^3 - (\alpha^2 + \beta^2 + \gamma^2)X.$$

If $u = \varphi(h)$, then as

$$(\mathrm{ad}\, u)(\varphi(e)) = [\varphi(h), \varphi(e)] = \varphi([h, e]) = 2\varphi(e)$$

ad u has 2 as an eigenvalue. Similarly, -2 must be an eigenvalue of ad u, and so is 0, so we need $\alpha^2 + \beta^2 + \gamma^2 = -4$. This suggests that we might try taking $u = 2iz$. Looking for eigenvectors of ad u, we find that $x + iy$ is in the 2-eigenspace and $x - iy$ is in the -2-eigenspace. We cannot immediately take these eigenvectors as the images of e and f because $[x + iy, x - iy] = -2iz$, whereas we would want $[\varphi(e), \varphi(f)] = \varphi([e, f]) = \varphi(h) = 2iz$. However, if we scale one of them by -1, we can also satisfy this requirement.

Therefore, if we define $\varphi : \mathsf{sl}(2, \mathbf{C}) \to L$ by $\varphi(h) = 2iz$, $\varphi(e) = x + iy$, and $\varphi(f) = -x + iy$, then by Exercise 1.9, φ will be one of the many possible isomorphisms.

1.16.   *Hint*: Try looking for a 2-dimensional algebra over F. Define the multiplication by a table specifying the products of basis elements.

Chapter 2

2.5. As z is a linear combination of commutators $[x,y]$ with $x,y \in L$, it is
sufficient to show that $\operatorname{tr} \operatorname{ad}[x,y] = 0$. But

$$\operatorname{tr} \operatorname{ad}[x,y] = \operatorname{tr}[\operatorname{ad} x, \operatorname{ad} y] = \operatorname{tr}(\operatorname{ad} x \circ \operatorname{ad} y - \operatorname{ad} y \circ \operatorname{ad} x) = 0.$$

2.8. Part (a) should not be found difficult, but as we use it many times, we
prove it here. Let $x_1, x_2 \in L_1$. We have $\varphi[x_1, x_2] = [\varphi x_1, \varphi x_2]$ so, by
linearity, $\varphi(L_1') \subseteq L_2'$. Now L_2' is spanned by commutators $[y_1, y_2]$ with
$y_1, y_2 \in L_2$, so to prove that we have equality, it is enough to note that
if $\varphi x_1 = y_1$ and $\varphi x_2 = y_2$, then $\varphi[x_1, x_2] = [y_1, y_2]$.

2.11. If $x \in \mathfrak{gl}_S(n, F)$, then $P^{-1}xP \in \mathfrak{gl}_T(n, F)$ since

$$(P^{-1}xP)^t T = P^t x^t P^{-t} T = P^t x^t S P = -P^t S x P = -T(P^{-1}xP).$$

We may therefore define a linear map

$$f : \mathfrak{gl}_S(n, F) \to \mathfrak{gl}_T(n, F), \quad f(x) = P^{-1}xP.$$

This map has inverse $y \mapsto PyP^{-1}$, so it defines a vector space isomorphism between $\mathfrak{gl}_S(n, F)$ and $\mathfrak{gl}_T(n, F)$. Moreover, f is a Lie algebra
homomorphism since if $x, y \in \mathfrak{gl}_S(n, F)$ then

$$\begin{aligned}
f([x,y]) &= P^{-1}(xy - yx)P \\
&= (P^{-1}xP)(P^{-1}yP) - (P^{-1}yP)(P^{-1}xP) \\
&= [P^{-1}xP, P^{-1}yP] \\
&= [f(x), f(y)].
\end{aligned}$$

See Exercise 4.10 for a related problem.

2.14.* (i) One approach is to argue that L is a Lie subalgebra of the Lie algebra
of all 3×3 matrices with entries in $\mathbf{R}[x,y]$.

(ii) The calculation shows that

$$[A(x^i, 0, 0), A(0, y^j, 0)] = A(0, 0, x^i y^i).$$

Hence L' is the subspace of all matrices of the form $A(0, 0, h(x, y))$ with
$h(x, y) \in \mathbf{R}[x, y]$.

(iii) Suppose $A(0, 0, x^2 + xy + y^2)$ is a commutator. Then, by part (ii),
there exist polynomials $f_1(x), f_2(x), g_1(y), g_2(y)$ such that

$$x^2 + xy + y^2 = f_1(x)g_2(y) - f_2(x)g_1(y).$$

Considering this as an equality of polynomials in x, with coefficients in the ring $\mathbf{R}[y]$, we see that either $f_1(x)$ has degree 2 and $f_2(x)$ has degree 0 or vice versa. In the first case, $g_2(y)$ must be a constant polynomial or else we get an unwanted $x^2 y^k$ term. Similarly, in the second case $g_1(y)$ must be constant, so there is no way we can obtain the xy term.

This exercise is based on Exercise 2.43 in Rotman's, *Introduction to the Theory of Groups* [20]. Rotman goes on to say that by taking a suitable ideal I in $\mathbf{R}[x, y]$ so that $\mathbf{R}[x, y]/I$ is finite, and replacing polynomials with suitable elements of the quotient ring, one can make the example finite-dimensional. For example one may take I to be the ideal generated by $x^3, y^3, x^2 y, x y^2$.

Chapter 3

3.2. We shall prove the "if" direction by defining an isomorphism φ between L_μ and $L_{\mu^{-1}}$. Let x_1, y_1, z_1 be a basis of L_μ chosen so that $\operatorname{ad} x_1$ acts on $L'_\mu = \operatorname{Span}\{y_1, z_1\}$ as the linear map with matrix

$$\begin{pmatrix} 1 & 0 \\ 0 & \mu \end{pmatrix}.$$

Let x_2, y_2, z_2 be the analogous basis of $L_{\mu^{-1}}$. We note that $\mu^{-1} \operatorname{ad} x_1$ has matrix

$$\begin{pmatrix} \mu^{-1} & 0 \\ 0 & 1 \end{pmatrix},$$

which is the matrix of $\operatorname{ad} x_2 : L'_{\mu^{-1}} \to L'_{\mu^{-1}}$, up to a swapping of the rows and columns. This suggests that we might define our isomorphism on basis elements by

$$\varphi(\mu^{-1} x_1) = x_2, \quad \varphi(y_1) = z_2, \quad \varphi(z_1) = y_2.$$

To check that this recipe really does define a Lie algebra isomorphism, it suffices to verify that $\varphi[x_1, y_1] = [\varphi x_1, \varphi y_1]$, $\varphi[x_1, z_1] = [\varphi x_1, \varphi z_1]$, and $\varphi[y_1, z_1] = [\varphi y_1, \varphi z_1]$; we leave this to the reader.

Now we tackle the harder "only if" direction. Suppose $\varphi : L_\mu \to L_\nu$ is an isomorphism. By Exercise 2.8(a), φ restricts to an isomorphism from L'_μ to L'_ν. As φ is surjective, we must have $\varphi x_1 = \alpha x_2 + w$ for some non-zero scalar α and some $w \in L'_\nu$. Let $v \in L'_\mu$. Calculating in L_μ gives

$$[\varphi x_1, \varphi v] = \varphi[x_1, v] = (\varphi \circ \operatorname{ad} x_1)v,$$

while calculating in L_ν gives

$$[\varphi x_1, \varphi v] = [\alpha x_2 + w, \varphi v] = \alpha(\mathrm{ad}\, x_2 \circ \varphi)v.$$

Thus $\varphi \circ \mathrm{ad}\, x_1 = \alpha\, \mathrm{ad}\, x_2 \circ \varphi = \mathrm{ad}(\alpha x_2) \circ \varphi$. As φ is an isomorphism, this says that the linear maps $\mathrm{ad}\, x_1 : L'_\mu \to L'_\mu$ and $\mathrm{ad}\, \alpha x_2 : L'_\nu \to L'_\nu$ are similar. In particular, they have the same sets of eigenvalues, hence $\{1, \mu\} = \{\alpha, \alpha\nu\}$.

There are now two possibilities: either $\alpha = 1$ and $\mu = \nu$, or $\alpha = \mu$ and $\mu\nu = 1$; that is, $\mu = \nu^{-1}$.

3.6. One can prove this by mimicking the approach used in §3.2.4 for the complex case. Step 1 and Step 2 remain valid over $\mathbf{R}$. If we still can find an element x of L such that $\mathrm{ad}\, x$ has a non-zero real eigenvalue, then Step 3 and Step 4 also go through unchanged.

However, it is now possible that for every non-zero $x \in L$ the linear map $\mathrm{ad}\, x$ has eigenvalues $0, \alpha, \bar{\alpha}$ where $\alpha \notin \mathbf{R}$. Suppose this is the case. Pick a non-zero $x \in L$ with eigenvalues $0, \alpha, \bar{\alpha}$. As before, $\mathrm{tr}\, \mathrm{ad}\, x = 0$, so α is purely imaginary. By scaling, we may assume that $\alpha = i$. Since $\mathrm{ad}\, x$ may be represented by the matrix

$$\begin{pmatrix} 0 & 0 & 0 \\ 0 & 0 & -1 \\ 0 & 1 & 0 \end{pmatrix},$$

there exist elements $y, z \in L$ such that $[x, y] = z$ and $[x, z] = -y$. One can now proceed as in Step 4 to determine the remaining structure constants and deduce that $L \cong \mathbf{R}^3_\wedge$.

It is easier to show that $\mathbf{R}^3_\wedge$ is not isomorphic to $\mathfrak{sl}(2, \mathbf{R})$: this is the content of Exercise 3.5.

Chapter 4

4.8. For (i) we take $a, b, c \in L$ and expand $0 = [[a, b + c], b + c]$ to get

$$[[a, b], c] = -[[a, c], b]. \qquad (\star)$$

The Jacobi identity states that for all $x, y, z \in L$

$$[[x, y], z] + [[y, z], x] + [[z, x], y] = 0.$$

By $(\star)$ the second term equals $-[[y, x], z] = [[x, y], z]$ and similarly the third term equals $[[x, y], z]$, so $3[[x, y], z] = 0$ for all $x, y, z \in L$. As L does not have characteristic 3, this implies that $L^3 = 0$.

Now assume that F has characteristic 3. The identity $(\star)$ implies that $[[x, y], z]$ is alternating; that is, permuting any two neighbouring entries changes its sign. In fact, $[[[x, y], z], t]$ is also alternating. This follows from $(\star)$ for the first three places, and also for the last, if one temporarily sets $w = [x, y]$.

We shall write $[x, y, z, t]$ for $[[[x, y], z], t]$. Using the alternating property and the Jacobi identity, we get

$$[[x, y], [z, t]] = [x, y, z, t] - [x, y, t, z] = 2[x, y, z, t]$$
$$= -[x, y, z, t].$$

This also gives

$$[[z, t], [x, y]] = -[z, t, x, y]$$

But $[z, t, x, y] = [x, y, z, t]$ (we swap an even number of times), and hence

$$[[x, y], [z, t]] = [[z, t], [x, y]] = [x, y, z, t].$$

Comparing this with the first equation we get $2[x, y, z, t] = 0$, and hence $L^4 = 0$.

Chapter 6

6.5. (ii) Suppose L is solvable. Then, by Lie's theorem, there is a basis of L in which all the matrices $\operatorname{ad} x$ for $x \in L$ are upper triangular. Hence, if $x \in L'$, then $\operatorname{ad} x$ is represented by a strictly upper triangular matrix. If follows that all the maps $\operatorname{ad} x$ for $x \in L'$ are nilpotent, so by Engel's Theorem L' is nilpotent.

The converse is easier, for if L' is nilpotent then L' is solvable, and so L is solvable.

Chapter 7

7.8. By Schur's Lemma (Lemma 7.13), z acts by scalar multiplication by some scalar λ on any finite-dimensional irreducible representation. But since $z = [f, g]$ is a commutator, the trace of the map representing z is zero. Hence $\lambda = 0$ and the representation is not faithful.

Alternatively, since the Heisenberg algebra is solvable, one can use Example 7.9(3).

Remark: The Heisenberg algebra models the position and momentum of particles in quantum mechanics. The fact that there are no faithful finite-dimensional irreducible representations (or even faithful irreducible representations by bounded linear operators on a Hilbert space) makes the mathematics involved in quantum theory quite a challenge.

7.9. To check that φ is a representation it is sufficient, since x, y form a basis of L, to check that $[\varphi(x), \varphi(y)] = \varphi([x, y]) = \varphi(x)$; this follows from

$$\left[\begin{pmatrix} 0 & 1 \\ 0 & 0 \end{pmatrix}, \begin{pmatrix} -1 & 1 \\ 0 & 0 \end{pmatrix} \right] = \begin{pmatrix} 0 & 1 \\ 0 & 0 \end{pmatrix}.$$

The matrices of $\operatorname{ad} x$ and $\operatorname{ad} y$ with respect to the basis x, y are

$$\operatorname{ad}(x) = \begin{pmatrix} 0 & 1 \\ 0 & 0 \end{pmatrix}, \quad \operatorname{ad}(y) = \begin{pmatrix} -1 & 0 \\ 0 & 0 \end{pmatrix}.$$

To show that the representations ad and φ are isomorphic, it is sufficient to find a linear map $\theta : \mathbf{C}^2 \to L$ such that

$$\theta \circ \operatorname{ad} x = \varphi(x) \circ \theta \quad \text{and} \quad \theta \circ \operatorname{ad} y = \varphi(y) \circ \theta$$

If θ has matrix P, then we need

$$P \begin{pmatrix} 0 & 1 \\ 0 & 0 \end{pmatrix} = \begin{pmatrix} 0 & 1 \\ 0 & 0 \end{pmatrix} P \quad \text{and} \quad P \begin{pmatrix} -1 & 0 \\ 0 & 0 \end{pmatrix} = \begin{pmatrix} -1 & 1 \\ 0 & 0 \end{pmatrix} P.$$

A small amount of calculation (starting with the first condition) now shows that we may take $P = \begin{pmatrix} 1 & 1 \\ 0 & 1 \end{pmatrix}$.

7.13. Suppose that W is an L-submodule of V. Then, in particular, W is an A-submodule of V. Each $a \in A$ acts diagonalisably on V and preserves the subspace W, hence W has a basis of a-eigenvectors. But commuting linear maps may be simultaneously diagonalised, so W has a basis of common eigenvectors for the elements of A.

In terms of weight spaces, this shows that if $\lambda_1, \ldots, \lambda_k \in A^\star$ are the weights of A appearing in the action of A on V so

$$V = \bigoplus_{i=1}^{k} V_{\lambda_i}$$

where $V_{\lambda_i} = \{ v \in V : a \cdot v = \lambda_i(a)v \text{ for all } v \in V \}$ is the weight space associated to λ_i, then

$$W = \bigoplus_{i=1}^{k} (W \cap V_{\lambda_i}).$$

Chapter 8

8.6.   (i) To show that $c : M \to M$ is a homomorphism of $\mathsf{sl}(2, \mathbf{C})$-modules, it is sufficient to show that it commutes with the action of e, f, and h. For $v \in M$, we have

$$
\begin{aligned}
e \cdot c(v) &= (efe + e^2 f + \tfrac{1}{2} eh^2) \cdot v \\
&= efe \cdot v + e(fe + h) \cdot v + \tfrac{1}{2}(he - 2e)h \cdot v \\
&= 2efe \cdot v + \tfrac{1}{2} heh \cdot v,
\end{aligned}
$$

whereas

$$
\begin{aligned}
c(e \cdot v) &= (efe + fe^2 + \tfrac{1}{2} h^2 e) \cdot v \\
&= efe \cdot v + (ef - h)e \cdot v + \tfrac{1}{2} h(eh + 2e) \cdot v \\
&= 2efe \cdot v + \tfrac{1}{2} heh \cdot v
\end{aligned}
$$

so c commutes with e. The proof for f is similar. For h, one checks that

$$
h(ef + fe) \cdot v = (ef + fe)h \cdot v
$$

using similar arguments.

(ii) Computing the action of c on the highest-weight vector $X^d \in V_d$ we get

$$
c(X^d) = (ef + fe + \tfrac{1}{2} h^2) \cdot X^d = dX^d + 0 + \tfrac{1}{2} d^2 X^d = \tfrac{1}{2} d(d + 2) X^d,
$$

so by Schur's Lemma c acts on V_d as multiplication by $\tfrac{1}{2} d(d + 2)$.

(iii) Since c is a module homomorphism, so is $(c - \lambda_i 1_M)^{m_i}$ for any $\lambda_i \in \mathbf{C}$, $m_i \geq 0$. The kernel of this map is then an $\mathsf{sl}(2, \mathbf{C})$-submodule.

(iv) By part (ii), we know that c acts on the irreducible submodule $U \cong V_d$ as multiplication by $\tfrac{1}{2} d(d + 2)$. By assumption, c has just one eigenvalue on the module M, so we must have $\lambda = \tfrac{1}{2} d(d + 2)$. Now suppose $W \cong V_{d'}$ is another irreducible submodule of M. Since λ is the unique eigenvalue of c on M, we must have

$$
\tfrac{1}{2} d(d + 2) = \tfrac{1}{2} d'(d' + 2),
$$

which implies that $d = d'$.

(v) Since N is a submodule of M, we can consider the action of c on the quotient space M/N,

$$
c(v + N) = (ef + fe + \tfrac{1}{2} h^2) \cdot (v + N).
$$

Since $M = \ker(c - \lambda 1_M)^m$ for some $m \geq 0$, we have

$$(c - \lambda 1_M)^m (v + N) = 0 \quad \text{for all } v \in M.$$

Hence c has just one eigenvalue, namely λ, on M/N. So, as in part (iv), any irreducible submodule of M/N is isomorphic to V_d.

Chapter 9

9.4. Take the standard basis e, f, h of $\mathsf{sl}(2, \mathbf{C})$. With respect to this basis,

$$\mathrm{ad}\, e = \begin{pmatrix} 0 & 0 & -2 \\ 0 & 0 & 0 \\ 0 & 1 & 0 \end{pmatrix}, \quad \mathrm{ad}\, f = \begin{pmatrix} 0 & 0 & 0 \\ 0 & 0 & 2 \\ -1 & 0 & 0 \end{pmatrix}, \quad \mathrm{ad}\, h = \begin{pmatrix} 2 & 0 & 0 \\ 0 & -2 & 0 \\ 0 & 0 & 0 \end{pmatrix}.$$

The matrix describing the Killing form is then

$$\begin{pmatrix} 0 & 4 & 0 \\ 4 & 0 & 0 \\ 0 & 0 & 8 \end{pmatrix}.$$

Since this matrix has rank 3, the Killing form is non-degenerate.

The Lie algebra $\mathsf{gl}(2, \mathbf{C})$ contains the identity matrix $I = \begin{pmatrix} 1 & 0 \\ 0 & 1 \end{pmatrix}$, which commutes with all elements of $\mathsf{gl}(2, \mathbf{C})$. Hence $\mathrm{ad}\, I = 0$ and $\kappa(I, x) = 0$ for each x, so $\mathsf{gl}(2, \mathbf{C})^{\perp}$ is non-zero.

9.6. We have already looked at $\mathsf{sl}(2, \mathbf{C})$. The Heisenberg algebra is nilpotent, so by Exercise 9.5 its Killing form is identically zero. The Lie algebra considered in §3.2.2 is a direct sum of the 2-dimensional non-abelian Lie algebra with the trivial Lie algebra, so its Killing form is known by Example 9.7.

Now let L be the 3-dimensional Lie algebra considered in §3.2.3. We saw that this Lie algebra has a basis x, y, z such that y, z span L' and the action of $\mathrm{ad}\, x$ on L' is represented by a matrix

$$\begin{pmatrix} 1 & 0 \\ 0 & \lambda \end{pmatrix}$$

for some $\lambda \in \mathbf{C}$. We find that $\kappa(b, c) = 0$ for all basis vectors b, c, with the only exception $\kappa(x, x) = 1 + \lambda^2$.

9.11. By Exercise 9.10, β and κ induce isomorphisms of L-modules $\theta_\beta : L \to L^\star$ and $\theta_\kappa : L \to L^\star$.

Consider the composite map $\theta_\kappa^{-1}\theta_\beta : L \to L$. As L is a simple Lie algebra, L is simple when considered as an L-module via the adjoint representation, so Schur's Lemma implies that there is a scalar $\lambda \in \mathbf{C}$ such that $\theta_\kappa^{-1}\theta_\beta = \lambda 1_L$, or equivalently $\theta_\kappa = \lambda\theta_\beta$. So by definition of the maps θ_κ and θ_β,

$$\kappa(x,y) = \theta_\kappa(x)(y) = \lambda\theta_\beta(x)(y) = \lambda\beta(x,y)$$

for all $x, y \in L$.

9.14. Let V be the given faithful representation. Using this representation, we may regard L as a subalgebra of $\mathsf{gl}(V)$. By Theorem 9.16, the abstract Jordan decomposition of $x \in L$ must agree with its Jordan decomposition as an element of $\mathsf{gl}(V)$. As x is diagonalisable, this implies that x is a semisimple element of L. It now follows from Theorem 9.16 that x acts diagonalisably in any representation of L.

9.15.* Let c be the Casimir operator on M; see Exercise 8.6. By this exercise, we may assume that c has only one eigenvalue on M. Moreover, this exercise shows that this eigenvalue is $\frac{1}{2}d(d+2)$, where d is the largest eigenvalue of h on M, and that any simple submodule of M, or more generally of a quotient of M, must be isomorphic to V_d.

The idea is to keep on taking off irreducible representations generated by highest-weight vectors. To make this argument explicit, we let U be a maximal completely reducible submodule of M, so U is a direct sum of irreducible representations each isomorphic to V_d. Our aim is to show that $U = M$.

Suppose that U is a proper submodule of M. In this case, M/U is non-zero, so it has a simple submodule, which must be isomorphic to V_d. Looking at the largest eigenvalue appearing in this submodule tells us that in M/U there is an h-eigenvector with eigenvalue d. As h acts diagonalisably on M this implies that there is an h-eigenvector $v \in M\backslash U$ such that $h \cdot v = dv$.

If $e \cdot v \neq 0$ then $e \cdot v$ would be an h-eigenvector with eigenvalue $d + 2$. Hence $e \cdot v = 0$ and v is a highest-weight vector. Let W be the submodule of M generated by w. By Corollary 8.6 (which says that a highest-weight vector in a finite-dimensional module generates an irreducible submodule), W is irreducible. As $w \notin U$, the irreducibility of W implies that $U \cap W = 0$, so $U \oplus W$ is a larger completely reducible submodule of M, a contradiction.

9.16. By Exercise 2.8, $\operatorname{ad}\theta(d)$ is diagonalisable. Suppose that $(\operatorname{ad}n)^m = 0$. Then, for all $y \in L_1$,

$$(\operatorname{ad}\theta(n))^m\theta(y) = [\theta(n), [\theta(n), \ldots, [\theta(n), \theta(y)]\ldots]]$$
$$= \theta([n, [n, \ldots, [n, y]\ldots]])$$
$$= \theta((\operatorname{ad}n)^m y) = 0.$$

As θ is surjective, this shows that $\operatorname{ad}\theta(n)$ is nilpotent. Moreover,

$$[\operatorname{ad}\theta(d), \operatorname{ad}\theta(n)] = \operatorname{ad}[\theta(d), \theta(n)] = \operatorname{ad}\theta([d, n]) = 0,$$

so $\operatorname{ad}\theta(d)$ and $\operatorname{ad}\theta(n)$ commute. By uniqueness of the Jordan decomposition, $\operatorname{ad}\theta(x) \in \mathfrak{gl}(L_2)$ has Jordan decomposition

$$\operatorname{ad}\theta(x) = \operatorname{ad}\theta(d) + \operatorname{ad}\theta(n).$$

The result now follows from the definition of the abstract Jordan decomposition.

Chapter 10

10.6. We may take $h_\alpha = e_{11} - e_{22}$ and $h_\beta = e_{22} - e_{33}$. If θ is the angle between α and β, then by Exercise 10.4,

$$4\cos^2(\theta) = 4\frac{(\alpha, \beta)}{(\beta, \beta)}\frac{(\beta, \alpha)}{(\alpha, \alpha)} = \alpha(h_\beta)\beta(h_\alpha) = -1 \times -1 = 1$$

and hence $\cos\theta = -\frac{1}{2}$.

The real subspace of H^* spanned by the roots looks like

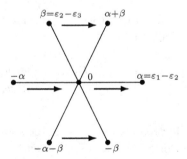

The arrows show the action of $\operatorname{ad}e_\alpha$ on the α root strings. (This action may be checked to be consistent with Proposition 10.10.)

10.12.   It is clear from Lemma 10.1(i) that N is invariant under e_α, and as a sum
of h_α-weight spaces, N is certainly invariant under h_α. The potential
problem comes when we apply f_α to elements of L_α. But by Lemma
10.6,
$$[f_\alpha, L_\alpha] \subseteq [L_{-\alpha}, L_\alpha] \subseteq \mathrm{Span}\,\{t_\alpha\} = \mathrm{Span}\,\{h_\alpha\},$$
so N is also closed under the action of f_α.

Consider the trace of $\mathrm{ad}\,h_\alpha : N \to N$. This must be zero because $\mathrm{ad}\,h_\alpha = [\mathrm{ad}\,e_\alpha, \mathrm{ad}\,f_\alpha]$ and so
$$\mathrm{tr}\,\mathrm{ad}\,h_\alpha = \mathrm{tr}\,(\mathrm{ad}\,e_\alpha\,\mathrm{ad}\,f_\alpha - \mathrm{ad}\,f_\alpha\,\mathrm{ad}\,e_\alpha) = 0.$$

On the other hand,
$$\mathrm{tr}\,\mathrm{ad}\,h_\alpha = -\alpha(h_\alpha) + 0 + \sum_{m \in \mathbf{N}} m\alpha(h_\alpha)\dim L_{m\alpha} = -2 + 2\sum_{m \in \mathbf{N}} m\dim L_{m\alpha}.$$

As $\dim L_\alpha \geq 1$, this shows that $\dim L_\alpha = 1$ and that the only positive
integral multiple of α that lies in $\varPhi$ is α. The negative multiples can be
dealt with similarly by considering $-\alpha$.

We finish much as in the proof of Proposition 10.9. Suppose that $(n+\tfrac{1}{2})\alpha$
is a root for some $n \in \mathbf{Z}$. The corresponding eigenvalue of h_α is $2n+1$,
which is odd. By the classification of Chapter 8, 1 must appear as an
eigenvalue of h_α on the root string module $\bigoplus_c L_{c\alpha}$. Therefore $\tfrac{1}{2}\alpha$ is a
root of L. But the first part of this solution shows that if α is a root
then 2α is not a root, so this is impossible.

Chapter 11

11.12.   We may assume without loss of generality that the U_i are pairwise distinct. We shall use induction on n, starting with $n = 2$. Assume for a
contradiction that $U_1 \cup U_2$ is a subspace. By the hypothesis, there exist
$u_1 \in U_1 \setminus U_2$ and $u_2 \in U_2 \setminus U_1$. Consider $u_1 + u_2$. If this element belongs
to U_1, then $u_2 = (u_1 + u_2) - u_1 \in U_1$, which is a contradiction. Similarly,
$u_1 + u_2 \notin U_2$, so $u_1 + u_2 \notin U_1 \cup U_2$.

For the inductive step, we assume the statement holds for $n-1$ distinct
proper subspaces of the same dimension. Assume for a contradiction
that $\bigcup_{i=1}^n U_i = U$ is a subspace. The subspace U_1 is not contained in
the union of $U_2, \ldots, U_n$, for this would mean that already $U = \bigcup_{i=2}^n U_i$,
contrary to the inductive hypothesis.

So there is some $u_1 \in U_1$ which does not lie in U_i for any $i \neq 1$. Similarly, there is some $u_2 \in U_2$ which does not lie in U_i for any $i \neq 2$. Now consider $u_1 + \lambda u_2$ for $\lambda \neq 0$. This lies in U, and hence it belongs to at least one of the U_i. Take two different non-zero scalars λ_1 and λ_2. Then the elements

$$u_1 + \lambda_1 u_2, \quad u_1 + \lambda_2 u_2$$

must lie in different spaces U_i; for instance, if they both belong to U_3 then so does their difference and hence $u_2 \in U_3$, contrary to the choice. The field contains $n - 1$ distinct non-zero scalars, so the corresponding $n - 1$ elements of the form $u_1 + \lambda_i u_2$ must lie in $n - 1$ distinct sets U_i. But since U_1 and U_2 are ruled out, there are only $n - 2$ sets available, so we have a contradiction.

Note that this proof only needs $|F| \geq n$.

11.9. Let $V_i = \mathrm{Span}\{b_j : j \neq i\}$ and let δ_j be a vector which spans $V_j^\perp$. Try $z = \sum r_i \delta_i$. We have by construction that $(z, b_j) = r_j(\delta_j, b_j)$ for each j. Now, $(\delta_j, b_j) \neq 0$ (otherwise b_j would be in V_j), so we can take $r_j = (\delta_j, b_j)$ and this z will do.

11.14. By restriction, we may view s_α and s_β as orthogonal transformations of the plane spanned by α and β. Let θ be the angle between α and β. We know that $s_\alpha s_\beta$ is a rotation, as it is an orthogonal transformation of the plane with determinant 1. To find its angle, we argue that if φ is the angle made between $s_\alpha s_\beta(\beta) = -\beta + \langle \beta, \alpha \rangle \alpha$ and β, then, since $s_\alpha s_\beta(\beta)$ has the same length as β,

$$2\cos\varphi = \langle s_\alpha s_\beta(\beta), \beta \rangle = -2 + \langle \beta, \alpha \rangle \langle \alpha, \beta \rangle = -2 + 4\cos^2\theta$$

so $\cos\varphi = 2\cos^2\theta - 1 = \cos 2\theta$. Thus $\varphi = \pm 2\theta$.

Using the bases $\{\alpha, \beta\}$ given in Example 11.6 for the 2-dimensional root systems, we get:

type	θ	order of $s_\alpha s_\beta$
$A_1 \times A_1$	π	2
A_2	$2\pi/3$	3
B_2	$3\pi/4$	4
G_2	$5\pi/6$	6

The hint given in the question now shows that the Weyl groups are $C_2 \times C_2$, D_6, D_8, and D_{12} respectively.

Chapter 12

12.3. (i) Consider the matrix describing κ, with respect to a basis compatible with the root space decomposition. It is block-diagonal. One diagonal block is κ restricted to $H \times H$. All other blocks are 2×2 blocks corresponding to pairs $\pm\alpha$. We must show that all these blocks are non-singular. That the block corresponding to H is non-degenerate follows from hypothesis (c). Let $\alpha \in \Phi$ and consider the corresponding block; it is of the form

$$\begin{pmatrix} 0 & \kappa(x_\alpha, x_{-\alpha}) \\ \kappa(x_\alpha, x_{-\alpha}) & 0 \end{pmatrix},$$

so we need only that this is non-zero. Let $h = [x_\alpha, x_{-\alpha}]$. We have

$$0 \neq \kappa(h,h) = \kappa(h, [x_\alpha, x_{-\alpha}]) = \kappa([h, x_\alpha], x_{-\alpha}) = \alpha(h)\kappa(x_\alpha, x_{-\alpha}),$$

as required.

(ii) Let $h \in H$ have diagonal entries $a_1, \ldots, a_{\ell+1}$, and similarly let $h' \in H$ have diagonal entries $a'_1, \ldots, a'_{\ell+1}$. The root space decomposition gives

$$\kappa(h, h') = \sum_{\alpha \in \Phi} \alpha(h)\alpha(h') = 2\sum_{i<j}(a_i - a_j)(a'_i - a'_j).$$

We write this as

$$2\sum_{i<j}(a_i a'_i + a_j a'_j) - 2\sum_{i<j}(a_i a'_j + a_j a'_i).$$

Consider $\sum_{i<j}(a_i a'_i + a_j a'_j)$. The total number of terms is $\ell(\ell+1)$. Each $a_k a'_k$ occurs the same number of times, so the first sum is equal to

$$2\ell \sum_i a_i a'_i = 2\ell \operatorname{tr}(hh').$$

Next consider the second sum. It can be written as

$$2\sum_i a_i \left(\sum_{j \neq i} a'_j\right) = 2\sum_i a_i(-a'_i) = -2\operatorname{tr}(hh').$$

Here we have used that h and h' have trace zero. Combining these, we have

$$\kappa(h, h') = 2(\ell+1)\operatorname{tr}(hh') = 2(\ell+1)\sum_i a_i a'_i.$$

Now $\sum a_i a'_i$ describes the usual inner product on $\mathbf{R}^{\ell+1}$ and hence κ is non-degenerate and $\kappa(h, h) = 0$ only if $h = 0$.

(iii) Let $L = \mathsf{so}(2\ell+1, \mathbf{C})$. Let h be a diagonal matrix of L with diagonal entries $0, a_1, \ldots, a_\ell, -a_1, \ldots, -a_\ell$, as in §12.3, and similarly let h' be another diagonal matrix. Making use of the table for the roots, we have

$$\kappa(h, h') = \sum_{\alpha \in \Phi} \alpha(h)\alpha(h')$$

$$= 2\sum_{i=1}^{\ell} a_i a_i' + 2\sum_{i<j}(a_i - a_j)(a_i' - a_j')$$

$$+ 2\sum_{i<j}(a_i + a_j)(a_i' + a_j')$$

$$= 2\sum_{i=1}^{\ell} a_i a_i' + 4\sum_{i<j}(a_i a_i' + a_j a_j').$$

Consider $\sum_{i<j}(a_i a_i' + a_j a_j')$. The counting argument we used for the case of $\mathsf{sl}(\ell+1, \mathbf{C})$ shows that this is equal to $(\ell-1)\sum_{i=1}^{\ell} a_i a_i'$. Hence

$$\kappa(h, h') = (4\ell - 2)\sum_{i=1}^{\ell} a_i a_i' = (2\ell - 1)\,\mathrm{tr}(hh').$$

This time $\sum a_i a_i'$ restricted to real matrices gives the usual inner product on $\mathbf{R}^\ell$. Hence κ is non-degenerate and $\kappa(h, h) = 0$ only if $h = 0$.

This essentially does all the calculations needed for $\mathsf{so}(2\ell, \mathbf{C})$ — the only difference is that the first term, $2\sum_{i=1}^{\ell} a_i a_i'$, does not appear. The calculations required for $\mathsf{sp}(2\ell, \mathbf{C})$ are similar.

12.4.* Let $I = \mathrm{rad}\,L$ and let $\varphi : L \to \mathsf{gl}(V)$ be the given faithful representation. We assume that $\mathrm{rad}\,L$ is non-zero. By Lie's Theorem, there is a vector $v \in V$ such that $\varphi(x)v \in \mathrm{Span}\{v\}$ for all $x \in I$. Let $\lambda \in I^\star$ be the associated weight. By the Invariance Lemma (Lemma 5.5), V_λ is an L-invariant subspace of V. But V is irreducible so $V_\lambda = V$. Hence $\mathrm{rad}\,L$ acts diagonalisably on V. Since V is faithful, this implies that $\mathrm{rad}\,L$ is abelian. Moreover, since $\ker \lambda \subseteq \ker \varphi = 0$, $\mathrm{rad}\,L$ must be 1-dimensional.

Let $\mathrm{rad}\,L = \mathrm{Span}\{z\}$. We may regard $\mathrm{Span}\{z\}$ as a 1-dimensional representation of the semisimple Lie algebra $L/\mathrm{rad}\,L$. The only 1-dimensional representation of a semisimple Lie algebra is the trivial representation, so $[x, z] = 0$ for all $x \in L$ and $\mathrm{rad}\,L$ is central.

To see that the radical splits off, we use Weyl's Theorem: regarding L as a representation of $L/Z(L)$ via the adjoint representation we see that the submodule $\mathrm{Span}\{z\}$ has an $L/Z(L)$-invariant complement. This complement will also be L-invariant, so we have $L = Z(L) \oplus M$ for some ideal M. Now $M \cong L/\mathrm{rad}\, L$ implies that $M' = M$ and hence $L' = M$.

(The converse also holds; namely, if $L = Z(L) \oplus L'$ with L' semisimple and $Z(L)$ 1-dimensional, then L has a faithful irreducible representation — see Exercise 15.7. The reader may wish to note that Lie algebras L with $\mathrm{rad}\, L = Z(L)$ are said to be *reductive*.)

Appendix A

16.6. (ii) Let W be the subspace of $\mathrm{Hom}(V, V)$ spanned by A. We may take $x_1, \ldots, x_m \in A$ so that

$$W = \mathrm{Span}\{x_1, \ldots, x_m\}.$$

By hypothesis, these maps commute so, by Lemma 16.7, there is a basis of V in which they are all represented by diagonal matrices. Since a linear combination of diagonal matrices is still diagonal, with respect to this basis every element of W, and hence every element of A, is represented by a diagonal matrix.

16.7. By hypothesis, d' commutes with x. By Lemma 16.8, we may express d as a polynomial in x, so d' and d commute. Similarly, n and n' commute.

By Lemma 16.7, there is a basis of V in which both d and d' are represented by diagonal matrices, so $d - d'$ is diagonalisable. On the other hand, since n and n' commute, $n - n'$ is nilpotent, so we must have $d - d' = n - n' = 0$; that is, $d = d'$ and $n = n'$.

Appendix D

19.1. Let $P = (p_{ij})$ and $Q = (q_{ij})$ for $1 \le i, j \le n$. We have $\sum_{i=1}^{n} p_{1i} q_{i1} = 1$, so there is a unique i_1 such that $p_{1i_1} = 1 = q_{1i_1}$. Since for $j \ne 1$ we have $\sum_{i=1}^{n} p_{1i} q_{ij} = 0$ and since $p_{1i_1} = 1$, it follows that $q_{i_1 j} = 0$; that is the row of Q with index i_1 is of the form

$$(1 \quad 0 \quad 0 \quad \cdots \quad 0).$$

Similarly, Q has a row of the form

$$(0 \quad 1 \quad 0 \quad \cdots \quad 0)$$

and so on. Hence Q is a permutation matrix, and since the inverse of a permutation matrix is also a permutation matrix, so is its inverse, P.

Bibliography

[1] E. Artin. *Geometric Algebra*. Wiley Classics Library. Wiley, New York, 1957, reprinted 1988.

[2] J. C. Baez. "The Octonions". *Bull. Amer. Math. Soc.*, 39:145–205, 2002.

[3] A. Baker. *Matrix Groups: An Introduction to Lie Group Theory*. Springer Undergraduate Mathematics Series. Springer, New York, 2003.

[4] T. S. Blyth and E. F. Robertson. *Basic Linear Algebra*. Springer Undergraduate Mathematics Series. Springer, New York, 1998.

[5] T. S. Blyth and E. F. Robertson. *Further Linear Algebra*. Springer Undergraduate Mathematics Series. Springer, New York, 2001.

[6] N. Bourbaki. *Lie Groups and Lie Algebras*. Springer, New York, 1989, 1998, 2002, 2005.

[7] R. W. Carter. *Simple Groups of Lie Type*. Wiley Classics Library. Wiley, New York, 1972, reprinted 1989.

[8] R. W. Carter. "Representations of the Monster". *Textos de Matemática, Série B, Univ. de Coimbra*, 33, 2002.

[9] J. Dixmier. *Enveloping Algebras*, volume 11 of American Mathematical Society Graduate Studies in Mathematics. American Mathematical Society, Providence, RI, 1977, reprinted 1996.

[10] W. Fulton and J. Harris. *Representation Theory, A First Course*, volume 129 of Graduate Texts in Mathematics. Springer, New York, 1991.

[11] P. R. Halmos. *Finite-Dimensional Vector Spaces*, 2nd edition, Undergraduate Texts in Mathematics. Springer, New York, 1958, reprinted 1987.

[12] T. Hawkins. "Wilhelm Killing and the structure of Lie algebras". *Arch. Hist. Exact. Sci.*, 26(2):127–192, 1982.

[13] J. E. Humphreys. *Linear Algebraic Groups*, volume 21 of Graduate Texts in Mathematics. Springer, New York, 1975.

[14] J. E. Humphreys. *Introduction to Lie Algebras and Representation Theory*, volume 9 of Graduate Texts in Mathematics. Springer, New York, 1978.

[15] N. Jacobson. *Lie Algebras*. Dover, New York, 1962, reprinted 1979.

[16] J. C. Jantzen. *Lectures on Quantum Groups*, volume 6 of American Mathematical Society Graduate Studies in Mathematics. American Mathematical Society, Providence, RI, 1996.

[17] J. C. Jantzen. "Representations of Lie Algebras in Prime Characteristic". In *Representation theories and algebraic geometry, NATO ASI, Series C, 514*. Kluwer, Dordrecht, 1998.

[18] V. Kac. *Infinite Dimensional Lie Algebras*, 3rd edition. Cambridge University Press, Cambridge, 1990.

[19] U. Ray. "Generalized Kac-Moody Lie Algebras and Related Topics". *Bull. Amer. Math. Soc.*, 38:1–42, 2001.

[20] J. J. Rotman. *An Introduction to the Theory of Groups*, 4th edition, volume 148 of Graduate Texts in Mathematics. Springer, New York, 1995.

[21] R. D. Schafer. *An Introduction to Nonassociative Algebras*. Dover, New York, 1996.

[22] H. Strade and R. Farnsteiner. *Modular Lie Algebras and Their Representations*, volume 116 of Marcel Dekker Textbooks and Monographs. Marcel Dekker, New York, 1988.

[23] A. Stubhaug. *The Mathematician Sophus Lie: It Was the Audacity of My Thinking*. Springer, Berlin, 2002.

[24] M. Vaughan-Lee. *The Restricted Burnside Problem*, 2nd edition, volume 8 of London Mathematical Society Monographs (Oxford Science Publications). Oxford University Press, Oxford, 1993.

Index